IUPAB Biophysics Series
sponsored by
The International Union of Pure and Applied Biophysics

Photosynthesis
Physical mechanisms and chemical patterns

IUPAB Biophysics Series
sponsored by
The International Union of Pure and Applied Biophysics
Editors:
Franklin Hutchinson
Yale University
Watson Fuller
University of Keele
Lorin J. Mullins
University of Maryland

Photosynthesis
Physical mechanisms and chemical patterns

RODERICK K. CLAYTON

Division of Biological Sciences
Cornell University, Ithaca, New York

CAMBRIDGE UNIVERSITY PRESS

CAMBRIDGE

LONDON   NEW YORK   NEW ROCHELLE

MELBOURNE   SYDNEY

Published by the Press Syndicate of the University of Cambridge
The Pitt Building, Trumpington Street, Cambridge CB2 1RP
32 East 57th Street, New York, NY 10022, USA
296 Beaconsfield Parade, Middle Park, Melbourne 3206, Australia

First published 1980

Printed in the United States of America
Typeset by Automated Composition Service Inc., Lancaster, Pennsylvania
Printed and bound by the Murray Printing Company, Westford, Massachusetts

Library of Congress Cataloging in Publication Data
Clayton, Roderick K.
Photosynthesis.
(IUPAB biophysics series; 4)
Includes bibliographies and index.
1. Photosynthesis – Research.   2. Photosynthesis.
I. Title.   II. Series: International Union for
Pure and Applied Biophysics. IUPAB biophysics
series; 4.
QK882.C58   581.1'3342   79-27543
ISBN 0 521 22300 8 hard covers
ISBN 0 521 29443 6 paperback

To B. J. Clayton, wife and colleague

CONTENTS

Contents

FOREWORD

The origins of this series were a number of discussions in the Education Committee and in the Council of the International Union of Pure and Applied Biophysics (IUPAB). The subject of the discussions was the writing of a textbook in biophysics; the driving force behind the talks was Professor Aharon Katchalsky, first while he was president of the Union, and later as the honorary vice-president.

As discussions progressed, the concept of a unified text was gradually replaced by that of a series of short inexpensive volumes, each devoted to a single topic. It was felt that this format would be more flexible and more suitable in light of the rapid advances in many areas of biophysics at present. Instructors can use the volumes in various combinations according to the needs of their courses; new volumes can be issued as new fields become important and as current texts become obsolete.

The International Union of Pure and Applied Biophysics was motivated to participate in the publication of such a series for two reasons. First, the Union is in a position to give advice on the need for texts in various areas. Second, and even more important, it can help in the search for authors who have both the specific scientific background and the breadth of vision needed to organize the knowledge in their fields in a useful and lasting way.

The texts are designed for students in the last years of the standard university curriculum and for Ph.D. and M.D. candidates taking advanced courses. They should also provide a suitable introduction for someone about to begin research in a particular field of biophysics. The Union is pleased to collaborate with the Cambridge University Press in making these texts available to students and scientists throughout the world.

Franklin Hutchinson, Yale University
Watson Fuller, University of Keele
Lorin J. Mullins, University of Maryland
Editors

PREFACE

The aim of this book is to introduce students of science to the methods and present state of research in photosynthesis. As befits a monograph on a topic in biophysics, physicochemical aspects of the subject are emphasized. The treatment of metabolic and physiological areas is confined to the earlier phases of ATP formation and carbon assimilation. The treatment of physical aspects is weighted heavily toward bacterial photosynthesis because the photosynthetic bacteria have afforded exceptional opportunities in elucidating physical mechanisms.

Part I describes major developments from about 1650 to 1960, emphasizing the chemical nature of photosynthesis and the roles of chlorophylls and other pigments. Part II reviews our present knowledge of the structures and components of photosynthetic tissues in relation to their function. Part III deals with the photochemistry of photosynthesis, and with the patterns of chemical events, principally electron and proton transfer, that follow the photochemistry. Part IV treats the relationships of electron and proton transport to ATP formation, and the metabolic patterns of carbon assimilation. The epilogue exposes major areas of confusion and ignorance and indicates potentially fruitful directions of research, including the development of photosynthetic systems for solar energy conversion.

This book can provide the framework for a course on photosynthesis suitable for undergraduate or postgraduate students. To meet this purpose it includes digressions into physics and chemistry, as needed for a basic understanding of the subject. These digressions can of course be passed over by the reader who is familiar with their content; they are mainly descriptive rather than analytical. I have tried to impart a knowledge of photosynthesis at the level of contemporary research; nevertheless, a student will be sufficiently prepared if he understands physics, chemistry, and biology at the level of introductory college courses for science majors. The reader will detect some redundancy in widely separated parts of the book. This is deliberate and is based on the premise that a cyclical return to some topics will help to consolidate an overall grasp of a subject as multifarious and complex as photosynthesis. The treatment is detailed and comprehensive so that students and mature investigators can visualize concretely what is required and involved in a career of research in photosynthesis.

Annotation has been restricted to a few references to specific experiments, plus a set of suggested readings at the end of each chapter. These readings are listed in the order of appearance of relevant material in the text.

I am indebted to Drs. W. L. Butler, R. E. McCarty, W. W. Parson, K. Sauer, and A. Vermeglio for valuable suggestions and critical comments on parts of this book.

R. K. C.

Part I

RESEARCH IN PHOTOSYNTHESIS: BASIC DEVELOPMENTS TO ABOUT 1960

Part I describes major developments in our understanding of photosynthesis during the past three centuries, especially from about 1930 to 1960. We shall trace the evolution of our views of two basic aspects of the subject: the primary photochemistry and the basic chemical patterns that surround this central process, and the roles of the pigments that absorb light energy and initiate the photochemistry.

The membranous structures that support photosynthesis in plants and in photosynthetic bacteria are described in Part II. It is satisfying to be able to visualize these structures while studying the physical and chemical processes that they mediate. To this end the reader is invited to examine the pictorial introduction to Part II while progressing through Part I.

The last chapter of Part I is an extended digression into molecular spectroscopy and optics applied to biology, to serve as preparation for the more detailed subsequent material.

1 The chemical nature of photosynthesis

1.1 Early history

An approximate overall equation for photosynthesis in green plants is

$$6CO_2 + 6H_2O \xrightarrow[\text{chlorophyll}]{\text{light}} C_6H_{12}O_6 + 6O_2 \qquad (1.1)$$

where $C_6H_{12}O_6$ represents a simple sugar. Let us see how the components of this reaction became recognized during the past three centuries.

About 1650 van Helmont grew a willow tree, starting with a 5-lb tree in 200 lb of sandy soil. Some 5 years later the tree weighed 570 lb, and the soil weighed 199 lb. Well before the enunciation of the law of conservation of matter (by Lomonosov in 1748 and Lavoisier in 1760), van Helmont guessed that most of the weight of the tree must have come from the water that had been added to the soil. He knew nothing of the role of carbon dioxide.

A century later Bonnet recorded that leaves submerged in water developed gas bubbles when placed in the sun. He was not likely to have been the first to see this, and he had no idea that a chemical process was involved. But the next three decades brought a remarkable growth in our knowledge of chemistry, and especially in the study of gases. In 1771 Joseph Priestley showed that a mouse could not live in a container with air that had been "burned out" by a candle flame, but that a spray of mint, perhaps chosen because of its fresh smell, restored the air so that after a few days a candle could burn again, or a mouse could live for a time. Thus Priestley showed that a plant could produce oxygen, or, in the language of his time, could dephlogisticate the air.

It remained for Ingenhousz to show, in 1779, that plants need their green parts and light in order to freshen the air, and that at night they spoil the air, just as a mouse or candle does (by respiration; formally the reverse, in the dark, of Reaction 1.1). Then in 1782 Senebier pointed out that plants need "fixed air," or carbon dioxide, to dephlogisticate (oxygenate) the air. Ingenhousz suggested in turn, in 1796, that this carbon dioxide was the source of all the organic matter in a plant. And finally in 1804 de Saussure,

3

aware of the law of conservation of matter, confirmed van Helmont's guess that most of the weight of a plant comes from water (and from CO_2).

By that time all the ingredients of Reaction 1.1 had become implicated in the life of plants. It remained only for Robert Mayer, who in parallel with Joule had developed the law of conservation of energy, to point out in 1845 that the energy taken up as sunlight was stored, in part, as chemical energy in the organic matter, represented in Reaction 1.1 as sugar. And it remained for organic chemists in the early twentieth century to characterize the essential green pigment, chlorophyll.

These chemists, notably R. Willstätter and A. Stoll, naturally tried to imagine how chlorophyll could mediate the process of photosynthesis. Guided only by notions of chemical plausibility, Willstätter and Stoll suggested that the reactants CO_2 and H_2O, which together form carbonic acid, are somehow bound to chlorophyll and rearranged when the chlorophyll absorbs light and goes into a state of greater energy. The rearrangement could create an unstable organoperoxide which would release oxygen:

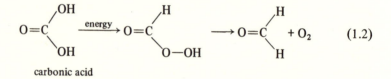

$$\underset{\text{carbonic acid}}{O=C\overset{\displaystyle\diagup OH}{\underset{\diagdown OH}{}}} \xrightarrow{\text{energy}} O=C\overset{\displaystyle\diagup H}{\underset{\diagdown O-OH}{}} \longrightarrow O=C\overset{\displaystyle\diagup H}{\underset{\diagdown H}{}} + O_2 \qquad (1.2)$$

In this view the first stable organic product of photosynthesis would be formaldehyde! All efforts to detect even a trace of formaldehyde in illuminated plant tissues failed. More significantly, this line of speculation suggested no new experiments or approaches to the problem. Then in the 1920s the microbiologist C. B. van Niel began to formulate a new, fruitful, and essentially correct view of the photochemical process of photosynthesis.

1.2   Photosynthesis as an oxidation–reduction process

C. B. van Niel was schooled in the study of comparative biochemical patterns among microbes. When he began to study photosynthesis in the 1920s, it was known that many types of bacteria can grow at the expense of light energy. The photosynthesis of these bacteria differs in several ways from that of green plants and algae:

1. The essential pigments, bacteriochlorophylls, are analogous to but chemically distinct from the chlorophylls of green plants.
2. No oxygen is evolved in bacterial photosynthesis.
3. In order for photosynthetic bacteria to assimilate carbon dioxide, they must be provided with a reducing substance. Among different

species of photosynthetic bacteria the list of such substances can include H_2, H_2S, and a variety of simple organic compounds such as alcohols and fatty acids. Generically this "oxidizable substrate" can be called H_2A. If it is an organic compound, it may serve both as a source of carbon (in place of CO_2) and as a reducing substance.

Van Niel noted that for many examples of bacterial photosynthesis the overall reaction could be approximated by

$$CO_2 + 2H_2A \xrightarrow[\text{bacteriochlorophyll}]{\text{light,}} (CH_2O) + H_2O + 2A \qquad (1.3)$$

where (CH_2O) represents stored organic matter. If the storage product is carbohydrate, (CH_2O) could denote one-sixth of a molecule of sugar. For green plants and algae the overall reaction (Reaction 1.1) could be rewritten, dividing by 6 and adding one H_2O to each side:

$$CO_2 + 2H_2O \xrightarrow[\text{chlorophyll}]{\text{light,}} (CH_2O) + H_2O + O_2 \qquad (1.4)$$

The similarity between Reactions 1.4 and 1.3 suggested to van Niel that green-plant photosynthesis is a special case in which water serves as the oxidizable substrate, and O_2 is the oxidation product. This gives green plants and algae a great advantage over the photosynthetic bacteria: Water is a ubiquitous and plentiful substrate.

Comparing these equations, van Niel proposed that photosynthesis should be seen as a coordinated pair of oxidizing and reducing processes: H_2A is oxidized to A, and CO_2 is reduced to (CH_2O). From these insights he developed a simple and elegant view of the photochemical part of photosynthesis: The light reaction, mediated by chlorophyll or bacteriochlorophyll (henceforth abbreviated Chl and Bchl), generates a pair of "primary" oxidizing and reducing entities. These in turn bring about the coordinated oxidation of H_2A and reduction of CO_2, respectively. Lacking specific knowledge, van Niel wrote the light reaction as a splitting of water into an oxidizing fragment denoted (OH) and a reducing fragment denoted (H):

$$H_2O \xrightarrow[\text{Chl or Bchl}]{\text{light,}} \begin{cases} \longrightarrow (H) \\ \\ \longrightarrow (OH) \end{cases} \qquad (1.5)$$

This was admittedly a mere formalism. There was no evidence that H_2O enters into the light reaction, and the primary oxidizing and reducing equivalents "(OH)" and "(H)" might just as well have been symbolized "+" and "−" or "X^+" and "Y^-." We shall retain van Niel's notation for the present, and come to grips later with the actual identities of the primary products.[1]

Reaction 1.5 could then be elaborated as follows:

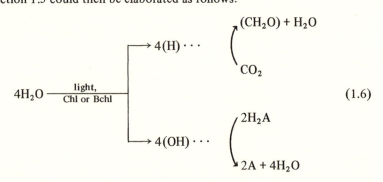

$$4H_2O \xrightarrow[\text{Chl or Bchl}]{\text{light,}} \begin{cases} \longrightarrow 4(H) \cdots \begin{cases} (CH_2O) + H_2O \\ \\ CO_2 \end{cases} \\ \\ \longrightarrow 4(OH) \cdots \begin{cases} 2H_2A \\ \\ 2A + 4H_2O \end{cases} \end{cases} \qquad (1.6)$$

Four reducing equivalents are needed to convert one CO_2 to (CH_2O), and four oxidizing equivalents to convert $2H_2A$ to $2A$ (or $2H_2O$ to O_2 in the case of green plants and algae). This view of photosynthesis had several implications:

1. The O_2 evolved by green plants comes entirely from H_2O and not from CO_2.
2. Neither the assimilation of CO_2 nor the evolution of O_2 (more generally, the oxidation of the substrate H_2A) is a part of the primary photochemistry.
3. The function of the substrate is to react with the primary oxidizing entity, so as to preserve an overall oxidation–reduction balance.
4. The photochemical act, which distinguishes photosynthesis from other ways of life, is a separation of oxidizing and reducing entities. These primary photoproducts carry some of the light energy into further chemical reactions, and coordinate the parallel processes of CO_2 reduction and H_2A oxidation.

We shall trace van Niel's formulation through a number of revisions. But his view of the photochemistry as a separation of oxidant and reductant, driven by light and mediated by Chl or Bchl, remains as a foundation of our current descriptions of photosynthesis.

1.3   Digression: qualitative aspects of oxidation and reduction

An understanding of oxidation and reduction is crucial to an understanding of photosynthesis. Oxidation and reduction are defined as the loss or gain of electrons, respectively:

$$X \underset{\text{reduction of } X^+}{\overset{\text{oxidation of } X}{\rightleftharpoons}} X^+ + e^- \qquad (1.7)$$

$$Y^- \underset{\text{reduction of } Y}{\overset{\text{oxidation of } Y^-}{\rightleftharpoons}} Y + e^- \qquad (1.8)$$

$$X + Y \underset{\substack{X^+ \text{ oxidizes } Y^-, \\ Y^- \text{ reduces } X^+}}{\overset{\substack{X \text{ reduces } Y, \\ Y \text{ oxidizes } X}}{\rightleftharpoons}} X^+ + Y^- \qquad (1.9)$$

An atom or molecule with high affinity for additional electrons acts as an oxidant; one that tends to release electrons acts as a reductant. These descriptions are of course relative to the surroundings of the atom or molecule. Thus, in Reaction 1.9 if the reaction proceeds spontaneously to the right, Y is a stronger oxidant than X; Y can take electrons away from X. Conversely if the state "X + Y" is more stable than the state "$X^+ + Y^-$," the reaction will go spontaneously to the left, and energy would be needed to drive it to the right. If photosynthesis begins with a reaction of the general form $X + Y \xrightarrow{\text{light, Chl}} X^+ + Y^-$ (a variation of van Niel's symbolism), one can imagine that part of the energy of light is used in driving the reaction, and becomes stored because "$X^+ + Y^-$" has more energy than "X + Y." We shall consider the energetics of oxidation and reduction at a quantitative level in Section 4.3.

In an aqueous environment, with H^+ ions (protons) ever present, the entities involved in oxidation and reduction may bind or release protons; for example,

$$Y + e^- \longrightarrow Y^- \qquad \text{followed by} \qquad Y^- + H^+ \longrightarrow YH \qquad (1.10)$$

If YH is more stable than the dissociated form $Y^- + H^+$ at the prevailing concentration of H^+, the reduction step in Reactions 1.10 will be followed rapidly by the protonation step, giving an overall reaction

$$Y + e^- + H^+ \longrightarrow YH \qquad (1.11)$$

This behavior is shown by many organic molecules at neutral pH. Thus the gain of an electron, if accompanied by protonation, is effectively the gain of an H atom. In such cases oxidation and reduction can be described in terms of H transfer instead of electron transfer:

$$A + BH \xrightarrow[\text{oxidation of B}]{\text{reduction of A,}} AH + B \qquad (1.12)$$

In this connection it is instructive to ask, for example, whether the conversion of two molecules of acetic acid to one of succinic acid,

$$2CH_3 \cdot COOH \longrightarrow HOOC \cdot CH_2 \cdot CH_2 \cdot COOH \qquad (1.13)$$

would represent a net oxidation or reduction, or neither. In this example the product has the same number of carbon and oxygen atoms as the reactants, but it has two H atoms fewer. The reaction involves a net oxidation, the removal of two H atoms (two electron equivalents). For a balanced reaction, 2H should be appended to the right side of Reaction 1.13.

As a second example, consider the relative levels of reduction of the two molecules $CH_2OH \cdot CHOH \cdot CH_2OH$ (glycerol) and $CH_3 \cdot CH_2 \cdot CH_2OH$ (propanol). The addition or removal of water from a molecule can be taken as neutral with respect to oxidation or reduction; so we can subtract H_2O as we like from these formulas, bringing them down to "skeleton" formulas of $CH \cdot C \cdot CH$ for glycerol and $CH_3 \cdot CH_2 \cdot CH$ for n-propanol. We can then see that n-propanol is four H atoms (or four electron equivalents) "more reduced" than glycerol. Another help in evaluating skeleton formulas is to say that the removal of an oxygen atom is equivalent to the insertion of two H atoms, because the same thing could be done by inserting two H atoms and then removing H_2O. The presence of an O atom thus counts for two oxidizing equivalents. Then by counting the number of O atoms (two oxidizing equivalents each) and H atoms (one reducing equivalent each), scaled to equal numbers of C atoms, one can rank the various metabolic products of photosynthesis as to their degree of reduction. This approach is useful in understanding the relationships between external nutrients and the products of their metabolic assimilation, by requiring an overall oxidation–reduction balance.

1.4   Modifications of van Niel's representation of photosynthesis

Some of the implications of van Niel's formulation (see the text following Reaction 1.6) received experimental scrutiny during the 1930s and 1940s. The idea that the O_2 evolved by green plants and algae comes not from CO_2 but from H_2O, as suggested by Reaction 1.4, could in principle be tested by means of the stable isotope ^{18}O. One could suspend algae in ^{18}O-labeled water with unlabeled CO_2, or vice versa, and examine the isotopic composition of the evolved O_2. Unfortunately this experiment can give ambiguous or misleading results for a variety of reasons. Oxygen can be exchanged between CO_2 and H_2O through the repeated formation and dissociation of carbonic acid, H_2CO_3. Exchange is generated by respiration (concurrent with photosynthesis), which makes new CO_2 out of O_2 and previously stored carbon compounds. There can be internal pools of O_2, CO_2, and H_2O that do not mix rapidly with the external medium. The chemistry of O_2 evolution, even if written simply as $2H_2O \longrightarrow O_2 + 4H^+ + 4e^-$, might involve CO_2 or bicarbonate in some secondary way. The first reports of this kind of experiment seemed to support van Niel's view, but with more repetitions and variations the results became ever more inconclusive, and there the matter rests.

Meanwhile two other implications of van Niel's model did receive clear support. Both the assimilation of CO_2 and the evolution of O_2 could be separated experimentally from the photochemical process in green plants and algae. R. Hill was making experiments with suspensions of chloroplasts, the green subcellular bodies that could be released from leaves by disrupting the tissue. He hoped to demonstrate complete photosynthesis in these suspensions, and he enjoyed a partial success. The chloroplasts did not convert CO_2 to sugar, but they did show a light-dependent evolution of O_2 if a suitable oxidizing agent had been added. The list of satisfactory "Hill oxidants" included Fe^{3+}, $Fe(CN)_6^{3-}$ (ferricyanide), and benzoquinone. These substances appeared to act as substitutes for CO_2, becoming reduced as a consequence of the light reaction. This *Hill reaction* (or chloroplast reaction, as Hill preferred to call it) seemed to show that the assimilation of CO_2 could be separated from the rest of photosynthesis. The evolution of O_2 could be coupled to the reduction of chemicals other than CO_2.

Just as CO_2 could be replaced by other reducible substances on the "reducing side" of van Niel's picture, so could the "oxidizing side" be modified, with the oxidation of hydrogen gas replacing the conversion of H_2O to O_2. This was shown by H. Gaffron in studies with certain green algae. If the algae were incubated in the dark in the absence of O_2, they acquired the ability to behave like photosynthetic bacteria. Using H_2 as the substrate "H_2A," they could assimilate CO_2 in the light without evolving O_2. This adaptation became lost in the light; the algae soon reverted to their normal O_2-evolving photosynthesis during illumination. But the potentiality had been shown: In these algae the evolution of O_2 could be separated from the rest of photosynthesis. The oxidizing product of the light reaction, "(OH)" in van Niel's early symbolism, could react with H_2O (giving O_2) *or* with H_2. The reducing product could drive the reduction of CO_2 *or* of a variety of other reducible substances, the so-called Hill reagents.

Ironically it was Gaffron who launched the strongest attack on van Niel's scheme in the 1930s and 1940s. Gaffron noticed that in some cases of bacterial photosynthesis the assimilation of carbon involved relatively little overall chemical change. Complex organic nutrients were assimilated, in a light-dependent reaction, with only minor alterations to convert them to storage products. If these cases were to be forced into van Niel's mold, one would imagine that the nutrient, acting as the substrate H_2A, is oxidized, perhaps even to the level of CO_2. These oxidation products are converted to storage products, drawing on the reducing entity formed by the light reaction. Gaffron argued that such a pattern is needlessly complicated, and suggested a simpler scheme of the general form

$$\text{light} + \text{Bchl} \longrightarrow \text{energy} \cdots \left\{ \begin{array}{l} \text{altered form for storage} \\ \\ \text{complex organic molecule} \end{array} \right. \quad (1.14)$$

This was before the word "energy" in Reaction 1.14 had been identified with adenosine triphosphate, ATP. When the chemical patterns of ATP formation began to emerge, in the hands of F. Lippman, O. Meyerhof, and others, it became possible to bring van Niel's picture of photosynthesis into harmony with that proposed by Gaffron. Initially van Niel held that any reaction between the primary photochemical products, such as a recombination of (H) and (OH) to regenerate H_2O, would be wasteful. The function of H_2A was to prevent such a reaction. But in the light of new knowledge one could visualize such a "back-reaction" as a means of transmuting and storing chemical energy. One could imagine that the primary reducing and oxidizing entities interact through a controlled sequence, with reducing equivalents (electrons or H atoms) cycling through a specific series of carriers from primary reductant to primary oxidant. The energy released in this exergonic process is captured, in part, through a coupled conversion of ADP to ATP (the mechanism of this energy coupling will be discussed in Chapter 10). Representing the photochemistry as the transfer of an electron from a donor P to an acceptor A, producing A^- and P^+ rather than van Niel's original (H) and (OH), one can write

$$\text{light} \rightsquigarrow \begin{pmatrix} \text{Chl} \\ \text{or} \\ \text{Bchl} \end{pmatrix} \quad \begin{matrix} A^- \\ \\ P^+ \end{matrix} \quad \sim \quad \begin{matrix} \text{ATP} + H_2O \\ \text{ADP} + H_3PO_4 \end{matrix} \tag{1.15}$$

The energy of ATP can then be used for all the needs of the living cell, including the metabolic conversions implied in Reaction 1.14. Among other things, the energy of ATP can drive electrons from a weaker reductant to a stronger one ("reductive dephosphorylation"). Then even the coupled oxidation of "H_2A" and reduction of CO_2 can be driven by the energy of ATP:

$$\text{light} \rightsquigarrow \begin{pmatrix} \text{Chl} \\ \text{or} \\ \text{Bchl} \end{pmatrix} \quad \begin{matrix} A^- \\ \\ P^+ \end{matrix} \quad \begin{matrix} \text{ATP} \\ \text{ADP} \end{matrix} \quad \begin{matrix} R^- \\ R \end{matrix} \cdots \begin{matrix} (CH_2O) \\ CO_2 \end{matrix}$$

$$\text{electrons from } H_2A$$

$$\tag{1.16}$$

In this scheme R^- is a strong reductant that can mediate CO_2 reduction.

This primitive representation will be amended and amplified as we proceed. For the present we should note that the overall chemical result is exactly the same as that of van Niel's original scheme. Light generates a separation of primary oxidant and reductant, H_2A is oxidized, and CO_2 is reduced. There is the added flexibility that ATP can be used for purposes other than the transfer of electrons from H_2A to R. For example, the scheme of Reaction 1.16 can embrace the less specific suggestion of Gaffron shown by Reaction 1.14. Reaction 1.16 also puts the role of H_2A in a new light. This substrate is not needed in the sense originally visualized by van Niel, to get rid of the primary oxidant and prevent a recombination between primary oxidant and reductant. Instead, H_2A can be drawn upon only as needed, to introduce reducing equivalents and maintain as overall redox balance between nutrients and storage products. Maximum metabolic flexibility is allowed if we accept both possibilities, the cyclic pattern of Reaction 1.16 and the noncyclic pattern of van Niel's initial formulation.

Noncyclic:

$$\text{light} \rightsquigarrow \begin{pmatrix} \text{Chl} \\ \text{or} \\ \text{Bchl} \end{pmatrix} \quad \begin{matrix} \text{A}\bar{\uparrow} \longrightarrow \text{electrons to reductant} \longrightarrow \text{reduction of } CO_2 \\ \\ \text{P}^+ \longleftarrow \text{electrons from } H_2A \end{matrix}$$

$$(1.17)$$

Cyclic:

$$\text{light} \rightsquigarrow \begin{pmatrix} \text{Chl} \\ \text{or} \\ \text{Bchl} \end{pmatrix} \quad \begin{matrix} \text{A}\bar{\uparrow} \\ \\ \text{P}^+ \end{matrix}$$

$$\sim \text{ATP} \begin{cases} \text{electrons to reductant} \longrightarrow \text{reduction of } CO_2 \\ \text{.Other uses of ATP including} \\ \quad \text{steps in the reduction of } CO_2 \\ \text{electrons from } H_2A \end{cases}$$

$$(1.18)$$

We shall see that both patterns have their place, and that the formation and the utilization of ATP are coupled to noncyclic patterns of electron flow as well as to cyclic patterns.

Gaffron's studies of the assimilation of organic nutrients by photosynthetic bacteria, which led to the viewpoint expressed by Reaction 1.14, were given more concrete form by Doudoroff, Stanier, and their collaborators about 1960. They found that organic compounds are stored in photosynthetic bacteria in two major forms, the counterparts of starch in green plants: a polymer of β-hydroxybutyric acid, $(-CH_2 \cdot CHOH \cdot CH_2 \cdot CO-)_n$, and the carbohydrate glycogen, $(C_6H_{10}O_5)_n$. If we subtract H_2O from these formulas until no oxygen atoms remain, and scale each of the resulting "skeleton formulas" to four carbon atoms, we have "C_4H_2" for the former and "C_4" for the latter, showing that poly-β-hydroxybutyric acid is more reduced than glycogen to the extent of two equivalents per four carbon atoms. Organic nutrients that are more oxidized than glycogen become stored mainly as glycogen, with CO_2 given off as needed to maintain an overall oxidation–reduction balance. Nutrients more reduced than β-hydroxybutyric acid are stored mainly as the polymer of this acid, with some CO_2 incorporation and/or H_2 evolution for oxidation-reduction balance. Nutrients at a level of reduction between these storage products are stored in both forms in suitable proportions. Probably the most convincing support of Gaffron's alternative to van Niel's original model is found when the substrate is β-hydroxybutyric acid. This is simply polymerized,

$$n\mathrm{CH_3 \cdot CHOH \cdot CH_2 \cdot COOH} \longrightarrow (-\mathrm{CH_2 \cdot CHOH \cdot CH_2 \cdot CO}-)_n + n\mathrm{H_2O}$$

with the help of a little energy provided by ATP.

On the other hand, when the source of carbon is solely CO_2 and the substrate is H_2, H_2S, or H_2O, van Niel's noncyclic formulation (Reaction 1.17) seems more direct and simple than the cyclic model (Reaction 1.18). This is the case with green plants and algae; a noncyclic transfer of electrons from H_2O ultimately to CO_2 is predominant. But with the bacteria, several lines of indirect evidence suggest that cyclic electron flow coupled to ATP formation is the principal mode of photosynthetic metabolism. The principal evidence, that in most cases the photosynthetic bacteria follow the cyclic pattern of Reaction 1.18, is that electrons are not transferred from substrates to strong reductants under conditions that interfere specifically with the formation of ATP. The main "strong reductants" that mediate CO_2 reduction are ferredoxins and pyridine nucleotides, which will be described later. The reduction of these substances, in illuminated cells or extracts of photosynthetic bacteria, is slowed markedly by poisons that interfere with the formation of ATP without inhibiting the photochemistry or the cycle of electron transport from A^- to P^+. The main energy-storing activity of most photosynthetic bacteria appears to be cyclic ATP formation.

The compounds labeled R/R^- in Reaction 1.16 and "reductant" in Reactions 1.17 and 1.18, which mediate the reduction of CO_2, are pyridine nucleotides: NADP in green plants and algae and (with less thorough documentation) NAD in photosynthetic bacteria. The structures of these molecules in their oxidized and reduced forms are shown in the next section. The present nomenclature, NAD (nicotinamide-adenine dinucleotide) and NADP (nicotinamide-adenine dinucleotide phosphate), replaces the earlier terms DPN (diphosphopyridine nucleotide) and TPN (triphosphopyridine nucleotide), and the still earlier terms Coenzyme I and Coenzyme II. Now that we have become used to the new terms, another revision is surely in order.

A final point about chemical patterns of photosynthesis, which should be anticipated now and will be expanded later, is that the photosynthesis of green plants and algae involves two different kinds of photochemical system acting in concert. In one system (Photosystem 1) the primary reductant is strong enough to reduce pyridine nucleotides, and the primary oxidant is relatively weak. The other system, Photosystem 2, generates a weak reductant and a strong oxidant, capable of taking electrons from water. The two photosystems operate in series:

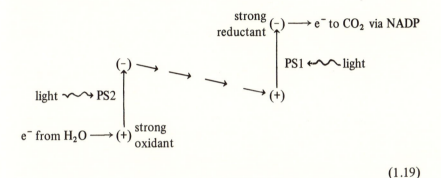

$$(1.19)$$

The weak reductant made by Photosystem 2 delivers electrons to neutralize the weak oxidant generated by Photosystem 1. The overall result is a non-cyclic transfer of electrons from water to NADP by means of two distinct photochemical systems of the general kind visualized by van Niel. As we develop the evidence for this pattern, commonly called the "Z scheme," we shall see how cyclic patterns of electron flow and the conservation of energy as ATP are superimposed on the basic outline. For photosynthetic bacteria, with two exceptions, there is no evidence for more than one kind of photochemical system. One exception is found in the unicellular blue-green algae, prokaryotic organisms that are now classified as bacteria. These *cyanobacteria* have the same chemical patterns of photosynthesis as green plants and other algae. The other exception is *Prochloron*, a genus of tropical prokaryotes with pigments and photochemical activities typical of green algae. For semantic convenience the cyanobacteria and *Prochloron* will be referred to as algae throughout this book. They are excluded when we speak of "bacterial photosynthesis."

As we have seen, van Niel's original formulation has undergone much revision and embellishment. In retrospect, his most remarkable and enduring contribution has been his prediction that the photochemistry of photosynthesis amounts to a separation of oxidizing and reducing entities. Modern thinking about the mechanism of photosynthesis began with this idea.

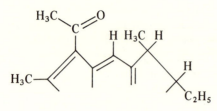

Bchl *a*

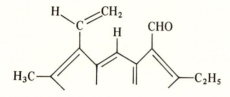

Chl *b*

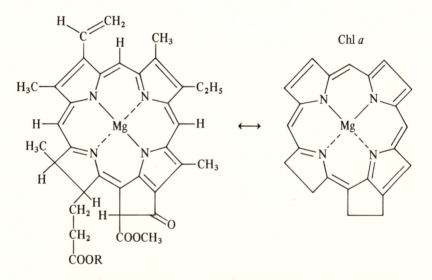

Chl *a*

Fig. 1.1. The structures of bacteriochlorophyll *a* and chlorophylls *a* and *b*. The presence of a carbon atom is implied at each unlabeled junction of bonds. In Chl *a* and *b*, but not in Bchl *a*, the pattern of alternating single and double bonds is in resonance with the one sketched at the right. The residue R is a long-chain hydrocarbon, $C_{20}H_{39}$ (phytyl) in Chl *a* and something similar in Bchl *a*.

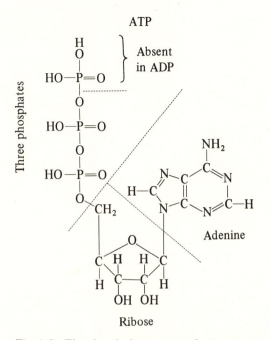

Fig. 1.2. The chemical structure of adenosine triphosphate (ATP).

1.5 Digression: structures of some molecules important in photosynthesis

In anticipation of the next chapter and for future reference, the structures of some chlorophylls and other molecules are presented here. Figure 1.1 shows the structures of Chl a, Chl b, and Bchl a.

The structures of adenosine di- and triphosphate, ADP and ATP, are shown in Fig. 1.2. The related structures of pyridine nucleotides, NAD and NADP, both oxidized and reduced forms, are shown in Fig. 1.3.

Quinones play major roles as electron carriers in photosynthesis. The structures of plastoquinone, found in plants, and of ubiquinone, found in bacteria, are shown in Fig. 1.4. The reduction of a quinone proceeds in two one-electron steps, illustrated for the simplest quinone, benzoquinone. The first step produces a semiquinone, and the second yields the fully reduced quinol, called hydroquinone in the case of benzoquinone. Each step may be attended by the binding of one or two protons, depending on the kinetics (the time available) and the concentration of H^+.

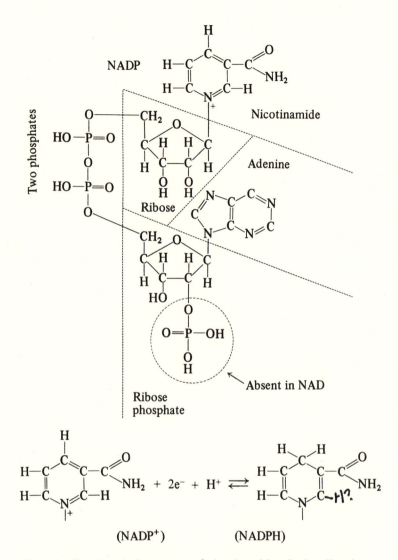

Fig. 1.3. The chemical structure of nicotinamide-adenine dinucleotide phosphate, NADP. Without the "odd" phosphate we have NAD. Note the similarity to ATP. Interconversion between the oxidized form NADP⁺ and the reduced form NADPH, which takes place on the nicotinamide part, is shown below.

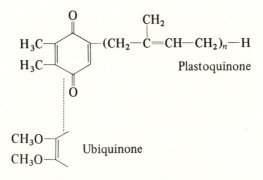

Plastoquinone

Ubiquinone

Benzoquinone | Anionic semiquinone | Protonated semiquinone

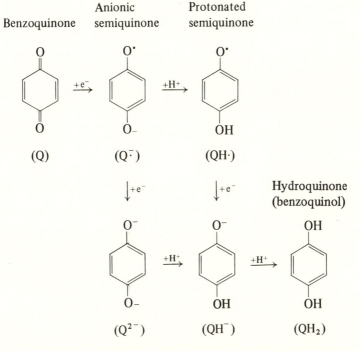

Hydroquinone (benzoquinol)

Fig. 1.4. Structures of quinones. Plastoquinones A, B, and so on, found in green plants, differ in the number (n) of isoprenoid groups in the side chain. The predominant plastoquinone A has $n = 9$. For the ubiquinones, found in photosynthetic bacteria, n is usually in the range 6 to 10. Major pathways in the reduction of quinones are illustrated for benzoquinone. Other forms, such as protonated quinone (QH^+) or doubly protonated semiquinone ($QH_2^{+\cdot}$), are possible, but are not likely to appear under physiological conditions.

SUGGESTED READINGS

Rabinowitch, E. (1971). An unfolding discovery. *Proc. Natl. Acad. Sci. U.S.* *68*, 2875-6.

Willstätter, R., and Stoll, A. (1913). *Untersuchungen über Chlorophyll.* Springer, Berlin.

Willstätter, R., and Stoll, A. (1918). *Untersuchungen über die Assimilation der Kohlensäure.* Springer, Berlin.

Fischer, H., and Orth, H. (1937). *Die Chemie des Pyrrols,* Vol. 2, Part 1. Akad. Verlagsges., Leipzig.

van Niel, C. B. (1941). The bacterial photosyntheses and their importance for the general problem of photosynthesis. *Adv. Enzymol. 1,* 263-328.

Hill, R. (1939). Oxygen production by isolated chloroplasts. *Proc. Roy. Soc. (London) B127,* 192-210.

Gaffron, H. (1940). Carbon dioxide reduction with molecular hydrogen in green algae. *Am. J. Bot. 27,* 273-83.

Gaffron, H. (1935). Über den Stoffwechsel der Purpurbakterien, II. *Biochem. Z. 275,* 301-19.

Lipmann, F. (1941). Metabolic generation and utilization of phosphate bond energy. *Adv. Enzymol. 1,* 99-162.

Stanier, R. Y. (1961). Photosynthetic mechanisms in bacteria and plants: Development of a unitary concept. *Bacteriol. Revs. 25,* 1-17.

2 The roles of chlorophylls and other pigments

2.1 Digression: light and molecules; excited states and energy transfer

Consider a pair of particles carrying equal but opposite electric charges. When viewed from a distance that is large compared to their separation, the pair of charges is called an electric dipole. Imagine that these charges are oscillating along the line joining them, as a sine function of time:

$$\overset{+}{-}, \ \pm, \ \overset{+}{+}, \ \overline{+}, \ \overset{-}{+}, \ \overline{+}, \text{etc.}$$

This movement generates around the charges a changing pattern of electric and magnetic forces called an electromagnetic wave, which is sketched in Fig. 2.1. At any instant the electric force, indicated by the vector $\overline{E}$, varies with distance from the dipole in the manner of a sine wave. The whole pattern moves away from the dipole with speed v. The magnetic force, not shown, is everywhere perpendicular and proportional to the electric force, and both are perpendicular to the axis of propagation. The distance from one crest of the sine wave to the next is the wavelength λ, and ν is the frequency with which successive crests pass a given point. The product of wavelength and frequency equals the speed of the wave:

$$\lambda \, (\text{cm}) \cdot \nu \, (\text{sec}^{-1}) = v \, (\text{cm sec}^{-1}) \tag{2.1}$$

In a vacuum the speed of an electromagnetic wave equals 3×10^{10} cm sec^{-1} and is denoted c. In a material medium such as water or glass the speed is less by a factor n, the index of refraction; $v = c/n$. For air, n is close to unity, and v is very nearly equal to c. The frequency of the wave equals the frequency of the dipole oscillation. Thus in a material medium the wavelength is reduced in proportion to the speed; see Equation 2.1.

Just as an oscillating dipole generates an electromagnetic wave, the forces represented by the wave can make another dipole oscillate. The wave thus carries energy from one dipole to another. Work is done in maintaining the oscillation of the first dipole, which radiates electromagnetic energy. The second dipole acquires some of this energy when set into motion. The density of energy in the wave is proportional to the square of the electric vector, E^2.

19

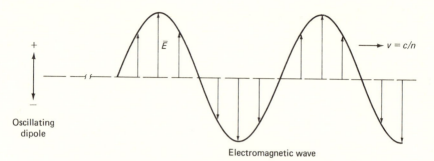

Fig. 2.1. An oscillating dipole is a source of an electromagnetic wave. The pattern of forces, electric ($\bar{E}$) and magnetic (not shown), comprising this wave is sinusoidal and moves away from the dipole with speed $v = c/n$. In a vacuum, and approximately in air, $v = c = 3 \times 10^{10}$ cm sec^{-1}. The wavelength is indicated by λ. (More details are given in the text.)

On an atomic or a molecular scale, electrons and positive nuclei can move relative to each other in the manner of an oscillating dipole. An incandescent body can be regarded as a collection of such oscillators, covering a spectrum of frequencies. The resulting radiation will induce dipole oscillations in other atoms or molecules, raising them to states of higher energy. The molecules of visual pigment in our eyes are especially responsive to oscillations in the narrow range of frequencies between 0.4×10^{15} sec^{-1} and 0.75×10^{15} sec^{-1}, corresponding to wavelengths from 4×10^{-5} to 7×10^{-5} cm. Electromagnetic waves in this part of the spectrum are therefore called visible light.

The unit of wavelength preferred by physicists is the angstrom (1 Å = 10^{-10} m); in these units the visible spectrum extends from 4000 Å (violet light) to 7000 Å (red). Biologists prefer the nanometer (1 nm = 10^{-9} m), with the visible going from 400 to 700 nm, perhaps because experimental data in biology seldom justify the use of four significant digits. The part of the spectrum adjacent to the visible on the short wave side, below 400 nm, is the ultraviolet; that on the long wave side is the infrared. The region from 700 nm to about 1100 nm is commonly called the near infrared. This region is important in photosynthesis because bacteriochlorophylls absorb strongly there.

At the turn of the century the classical picture of oscillating dipoles and electromagnetic waves encountered great difficulties when applied on the atomic or molecular scale. These difficulties had to do with the way that the electromagnetic wave interacts with matter, and they gave rise to quantum theory. The difficulties were:

1. The theoretical distribution of energies emitted by an incandescent body at various frequencies was incorrect. When this emission spectrum was integrated to compute the total energy emitted per unit of time, the answer was infinity!

2. Light falling on a metal surface can eject electrons from the metal (the photoelectric effect). The maximum energy of an ejected electron was found to be related to the frequency of the light and not to the intensity of the beam: $E_{max} = h\nu - w$, where w is a constant (the work function) characteristic of the metal, and h is a universal natural constant (see following paragraphs).

3. In Rutherford's classical picture of an atom, electrons move in orbits around the nucleus. Such an atom should be unstable. The orbital motion can be regarded as two mutually perpendicular dipole oscillations, $90°$ out of phase with each other. Therefore this "planetary" system should radiate electromagnetic energy. The atom should lose energy, and the electrons should spiral into the nucleus.

4. Atoms, when heated or given energy in an electric discharge, emit light at a number of highly specific wavelengths (line spectra). The distinctive set of wavelengths in the emission spectrum of any given type of atom could be tabulated, and numerical relationships between these wavelengths could be discerned, but no physical sense could be made of this.

To meet the first of these difficulties Max Planck advanced a radical hypothesis: The electromagnetic field can exchange energy with matter only in indivisible packets or quanta of energy given by

$$E = h\nu \tag{2.2}$$

(this E should not be confused with the magnitude of the electric vector $\overline{E}$). The energy of a quantum is proportional to the frequency of the electromagnetic wave. The constant of proportionality h (Planck's constant) equals 6.6×10^{-34} joule sec when the energy is expressed in joules. When the theoretical emission spectrum of an incandescent body was revised to accommodate this rule, it came into agreement with experimental observation.

Einstein showed that in the photoelectric effect, the curious relationship between maximum electron energy and frequency of the light could be expected from the same rule using the same numerical constant h.[1]

The third difficulty, the instability of a planetary atom, was erased by a new model of the atom. N. Bohr postulated that an atom can exist only in one or another of a distinctive (quantized) set of energy states. It can gain or lose energy only in quanta of magnitudes equal to the differences between pairs of these states. The state of lowest energy can be called the ground state, and the others excited states. An atom can go from one state to another by absorbing or emitting a quantum of radiant energy, with the frequency of the radiation matching the change of energy according to Equation 2.2:

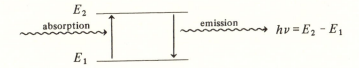

The energies of states allowed by the mathematical form of Bohr's postulate were in harmony with the distinctive set of wavelengths or frequencies of light observed to be absorbed and emitted by atoms, to a first approximation.

Planck's hypothesis and Bohr's postulate represented the beginnings of quantum theory. The quantization of energy had profound implications for the philosophy of observation and measurement. One cannot observe a physical system without perturbing it to a finite extent because the medium of observation (such as light) is itself quantized. Analysis of this situation led to the uncertainty principle: one cannot know with arbitrary precision both the position and the momentum of an object. The more accurately we know the present position of an object, the less we can say about its future position because of our uncertainty of its momentum. At best we can formulate a set of present and future probabilities to describe a physical situation. Instead of imagining that electrons move in specific orbits around an atomic nucleus, we specify the relative probabilities of finding the electron in one place or another relative to the nucleus.

The probability distribution of a particle, such as an electron, is given by a function of the space and time coordinates denoted ψ (the actual probability density, per unit volume of space, equals ψ^2). Because the distribution has a wavelike nature, ψ is called the wave function. Thus the particles of matter have a wavelike property, just as the radiation described as an electromagnetic wave has a particulate (quantum) nature.

The probability that an atom will absorb light and go to a state of higher energy depends on three factors. First, the light must be of the correct frequency so that a quantum matches the change of energy of the atom, $h\nu = \Delta E$. Second, the wave functions of the electrons in the excited state must occupy more or less the same regions as in the ground state: The greater the overlap is between excited- and ground-state wave functions, the more probable the transition. Finally, the redistribution of charge that occurs during the transition, reflected in the alteration of wave functions, must have the character of a transient electric dipole oscillation of frequency ν. These interrelated requirements determine whether an atom is likely to interact with radiation of a given frequency. They apply also to "downward" transitions, in which the redistribution of electrons amounts to a transient dipole oscillation attended by the emission of a quantum of the appropriate frequency. Transitions are "forbidden" if the redistribution of charge has no dipole character; an example is the spherically symmetric pulsation of electron distribution that would occur if a hydrogen atom went from its ground state to the next higher state ($1s \rightarrow 2s$ in the notation of spectroscopists) or vice versa.

When atoms come together to form molecules, a great variety of new interactions become possible. Electrons can interact with more than one nucleus, and the nuclei (or groups of nuclei) can move relative to each other, vibrating and rotating. The number of distinct energy states allowed by quantum theory is enormously greater than in the separate atoms. A single electronic state, characterized by certain basic electron distributions, is split into many substates corresponding to different quantized energies of vibration and rotation of nuclei. A transition between an electronic ground state and an excited state can therefore span a greater or lesser energy gap, depending on the particular substates at the start and end of the transition:[2]

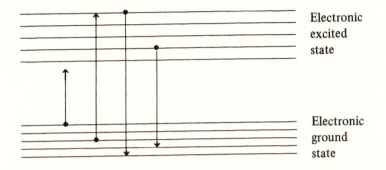

Electronic
excited
state

Electronic
ground
state

As a result, the frequency of light absorbed or emitted in a transition is not restricted to a single well-defined value; a range of frequencies is possible.

In a collection of molecules the energies will be distributed over substates as governed by thermal equilibrium, and at room temperature most will be in substates near the bottom of the ground state. This will be the most likely starting point of an upward transition. The most likely terminus is near the center of an excited state, where the density of substates is greatest. Once in an excited state, a molecule can emit a quantum and return to the ground state; this phenomenon is called fluorescence. Given enough time (more than about 10^{-11} sec), the excited molecule will have reached thermal equilibrium with its surroundings, and the downward transition will originate most probably from a substate near the bottom of the excited-state manifold. The most likely terminus is near the center of the ground-state manifold. These points are illustrated in Fig. 2.2. The changes in energy for the most probable upward and downward transitions are labeled $\Delta E_m{}^a$ (absorption) and $\Delta E_m{}^f$ (fluorescence). Note that $\Delta E_m{}^a > \Delta E_m{}^f$. If the most probable upward transition is followed by the most probable downward one, energy is lost as heat at two points in the cycle, after each of the major transitions, owing to thermal equilibration with the surroundings. The energy losses are shown in Fig. 2.2 by wavy arrows, and can be regarded as a succession of small transitions between closely spaced substates.

These remarks can be translated into spectra: plots of the intensity of absorption or fluorescence versus the frequency or wavelength of the light absorbed or emitted. Imagine that in Fig. 2.2 the substates are so close together that they cannot be resolved individually. The corresponding absorp-

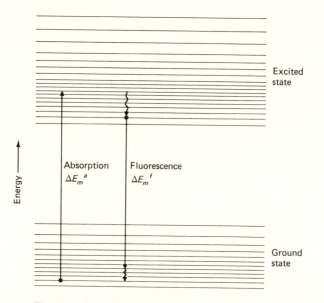

Fig. 2.2. Transitions between ground and excited electronic states of a molecule. The changes of energy for the most probable upward and downward transitions (absorption of light and fluorescence respectively) are labeled $\Delta E_m{}^a$ and $\Delta E_m{}^f$. The wavy arrows indicate losses of energy accompanying thermal equilibration with the surroundings.

tion and fluorescence spectra then show broad bands as sketched in Fig. 2.3, rather than the sharp "line spectra" of atoms. The ordinate in Fig. 2.3, intensity of absorption or fluorescence, will be defined quantitatively later in this book. On a frequency plot the absorption and fluorescence bands have their maxima where $h\nu = \Delta E_m{}^a$ and $\Delta E_m{}^f$, respectively. The peak of absorption comes at a higher frequency than the fluorescence peak; this is called the Stokes shift. On a wavelength plot the absorption peak comes at a lesser wavelength; energy is related inversely to wavelength: From Equation 2.1, $\Delta E = h\nu = hc/\lambda$ (if λ is measured in air, $\nu \approx c$). Physicists and chemists usually plot spectra in terms of frequency because frequency is proportional to energy. Biologists tend to use wavelength, apparently out of tradition. Wave number, denoted k and defined as the reciprocal of the wavelength in centimeters, is often used in place of frequency: $k = 1/\lambda = \nu/c$.

A molecule may be promoted to a higher excited state by absorbing a quantum of sufficient energy, as shown in Fig. 2.4. In that case it will probably relax to the lowest excited state before emitting a fluorescent quantum and returning to the ground state. This relaxation from higher excited states, due to thermal equilibration, approaches completion in about 10^{-11} sec. It is

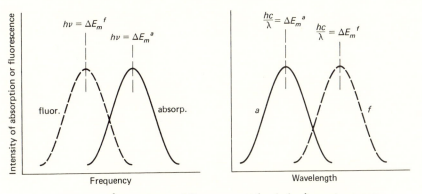

Fig. 2.3. Absorption and fluorescence (emission) spectra corresponding to the diagram of Fig. 2.2. (See discussion in the text.)

rapid because the higher excited states are well connected by overlapping substates that provide pathways for the process. From the lowest excited state to the ground state there is a gap where connecting substates are sparse, and a molecule is likely to stay in the lowest excited state for about 10^{-8} sec before giving up a quantum of fluorescence and returning to the ground state. Direct return from a higher excited state to the ground state, by emission of an energetic quantum of fluorescence, is relatively rare, requiring about 10^{-8} sec as contrasted with the 10^{-11} sec needed for relaxation. Any fluorescence from a higher state is therefore much weaker (by about $10^{-11}/10^{-8}$, or $1/1000$) than the fluorescence from the lowest excited state. The absorption spectrum will show two bands (see Fig. 2.4), but the fluorescence spectrum will show only one, corresponding to the return from the lowest excited state to ground.

We shall see that in photosynthesis the events that precede photochemistry are slow compared to thermal relaxation from higher states. When Chl is excited by light, it has nearly always relaxed to the lowest excited state before the energy is used for photochemistry. It follows that quanta of blue light, producing higher excited states in Chl, are no more useful for photosynthesis than the less energetic red quanta that generate the lowest excited state.

An important process in photosynthesis, as we shall see, is the transfer of excitation energy from one molecule of Chl to another. This transfer is not by the trivial process in which one molecule emits a quantum and another absorbs it. The mechanism is more intimate; it resembles the interaction of mechanically coupled pendulums, and is called resonance transfer. The dipole oscillation associated with de-excitation in one molecule is coupled at close range to a sympathetic oscillation in a neighboring molecule, causing excitation of the latter. The first molecule donates its energy directly to the second,

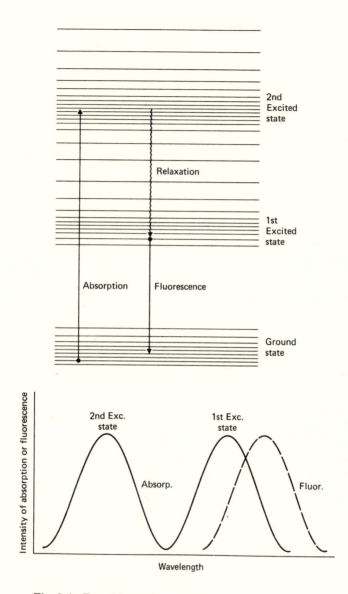

Fig. 2.4. Transitions of a molecule among higher excited states, and the corresponding spectra of absorption and fluorescence. (See discussion in the text.)

without the intervention of fluorescence or absorption in the usual sense. If the two molecules are different, as when Chl *b* transfers energy to Chl *a*, the energy lost by one must equal the energy gained by the other:

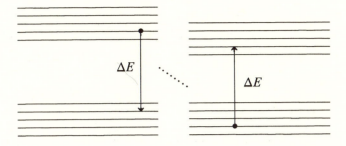

Energy transfer among identical molecules, as in a dense bed of Chl *a*, can be described in another way. The set of molecules can be treated as a single supermolecule; an excited state becomes a property of the supermolecule and not of any individual molecule. The quantum of excitation energy is thus delocalized over the whole set. The choice between these alternative descriptions, a hopping of energy from one molecule to the next or a delocalization of the energy over many, depends on the strength of the electric dipole interaction between neighboring molecules. With stronger interaction the "delocalized" model becomes more appropriate. For the chlorophylls in living tissues the strength of interaction seems to fall in a middle ground between these extremes. We shall use the "hopping" description of resonance transfer, mainly out of verbal convenience.

2.2   Two functions of chlorophyll: light harvesting and photochemistry

With one special exception,[3] every organism that lives by photosynthesis contains some sort of chlorophyll. We shall see that some of the Chl is involved directly in the primary photochemistry, but the greater part acts only to absorb light and transfer the excitation energy to the sites of photochemistry.

In the past two decades it has become established that in every known case, for green plants, algae, and photosynthetic bacteria, the primary photochemistry involves the transfer of an electron from Chl or Bchl to some sort of electron acceptor. The primary oxidizing and reducing entities, denoted P^+ and A^- in Reaction 1.15, are actually oxidized Chl (or Bchl) and reduced acceptor. The ways that various chlorophylls and electron acceptors are organized in photochemical reaction centers will be a major subject of this book;

for the present we shall use a simplified and nonspecific representation of the photochemistry:

$$
\begin{array}{c}
\begin{array}{ccccc}
\text{A} & \xrightarrow{h\nu} & \text{A} & \longrightarrow & \text{A}^{-} \\
\text{Chl} & & \text{Chl*} & & \text{Chl}^{+}
\end{array} \\
\left(\begin{array}{l}
\text{e}^{-} \text{ from A}^{-} \text{ to} \\
\text{secondary acceptor;} \\
\text{e}^{-} \text{ from donor to Chl}^{+}
\end{array}\right)
\end{array}
\tag{2.3}
$$

Chl is promoted to an excited state (Chl*) by absorbing a quantum of light ($h\nu$) or receiving it by transfer from other molecules of Chl. The excited Chl* donates an electron to the acceptor, and the products Chl$^+$ and A$^-$ act as starting points for subsequent electron transfer processes that generate ATP and strong reductants. These secondary electron transfers restore the system to the state Chl·A, so that it can perform its photochemistry again. In healthy photosynthetic tissues the entire cycle takes a few milliseconds, affording a turnover rate of the order of 100 times per second.

Imagine a primordial photosynthetic microbe containing photochemical systems of this kind, Chl·A, and no other Chl or accessory pigments. This creature was growing in shaded daylight, avoiding the direct flux of solar ultraviolet that penetrated the primitive atmosphere. One can compute (see Section 4.4) that each molecule of Chl would absorb no more than about one quantum per second under such conditions. The photochemistry of such an organism would have been grossly underdriven, when compared with a turn-over potentiality of 100 per second. It would have been a simple evolutionary maneuver to surround each photochemical center with additional molecules of Chl,

Chl Chl
Chl Chl·A
Chl Chl

close enough that a quantum absorbed by any of these Chls would have had a high probability of being transferred to the photochemically active Chl·A before being dissipated as fluorescence or heat. Such an organism, having amplified its ability to collect light energy, would have had a great advantage over its ancestors by making full use of the rapid kinetics of the dark reactions.

Contemporary photosynthetic organisms do indeed contain from about 50 to more than 1000 molecules of "light-harvesting" or "antenna" Chl, plus other light-gathering pigments, for each photochemical reaction center. This arrangement was first appreciated by R. Emerson and W. Arnold in the early 1930s as a result of their studies of photosynthesis in intermittent light. Using suspensions of the green alga *Chlorella*, a favorite research organism at the

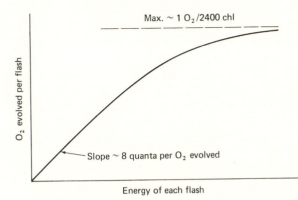

Fig. 2.5. Emerson and Arnold's experiment showing the existence of photosynthetic units; systems of photochemical centers plus antenna (light-harvesting) Chl. *Chlorella* algae were exposed to a succession of brief flashes of light separated by dark intervals greater than 0.04 sec. (Further description and analysis are given in the text.)

time, they undertook to measure the dynamics of "light" and "dark" processes. They exposed the algae to a succession of brief, intense flashes of light: so intense that every photochemical center would be activated, and, it was hoped, so brief (about 10^{-5} sec, from the spark of a capacitor discharge) that each photochemical center could react just once. They measured the amount of oxygen evolved per flash, as a function of the interval between flashes. With intervals greater than about 0.04 sec the yield per flash was maximal, signifying that each photochemical center, plus the chemical machinery for evolving oxygen, had had enough time to complete its cycle after one flash and be ready to make full use of the next.

Emerson and Arnold then measured the yield of oxygen as a function of the energy of the flash, keeping the dark interval between flashes greater than 0.04 sec. The results are outlined in Fig. 2.5. With weak flashes the yield corresponded to 1 O_2/about 8 quanta absorbed by the algae. Strong flashes, sufficient to excite every molecule of Chl at least once, gave a maximal yield of about 1 O_2/2400 Chl. This was a suprise at the time. If 8 quanta absorbed by Chl can yield 1 O_2, a maximal yield of 1 O_2/8 Chl might have been expected from a flash that excites every Chl. Emerson and Arnold reasoned that 2400 molecules of Chl cooperate in serving a single machine for making O_2, so that 8 quanta absorbed anywhere in the set of 2400 Chls can collectively provide for the evolution of 1 O_2. Additional excitations in this set during the brief time of the flash are wasted. Emerson and Arnold called this set of Chls, plus the associated "dark" chemical machinery, a photosynthetic unit. But if the primary photochemistry involves single quanta driving single electron transfers, the interpretation could be rephrased. One quantum absorbed by any Chl in a set of 300 (2400 ÷ 8) can produce one photochemical event, and the fruits of eight such photochemical electron transfers are needed for the evolution of one O_2.

Arnold and H. Kohn developed the latter viewpoint by analyzing the curve sketched in Fig. 2.5. The onset of saturation, where the yield of O_2 begins to level off, will occur with weaker flashes if a greater number of Chl molecules can pool their excitation energy to serve a single photochemical center. The actual curve corresponded to an effective absorption cross section, for a single "one-electron" photochemical center plus the associated "antenna" Chl, equal to that of about 300 Chls.

The concept of a photosynthetic unit first advanced by Emerson and Arnold became refined in subsequent years. It should not necessarily be visualized as a distinct package of antenna pigment plus reaction center, isolated from neighboring photosynthetic units. Perhaps the antenna pigments form a matrix studded with reaction centers in the ratio 300 Chl to one reaction center. These contrasting views have been called the "puddle" and "lake" models, respectively. They can be distinguished by the kinetics of utilization of excitation quanta.

Referring to Reaction 2.3, only the state Chl·A of the photochemical reaction center is active. Other states such as $Chl^+ \cdot A^-$, $Chl^+ \cdot A$, or $Chl \cdot A^-$ are not ready to perform photochemistry; they are inactive. Now, in the lake model a quantum can visit more than one reaction center during its migration in the antenna. Then if the quantum encounters an inactive center, it still has a chance to visit another reaction center that is active. In the puddle model this option is missing. The distinction can be approached experimentally by measuring the efficiency with which quanta are utilized during continuous illumination, when some of the reaction centers are inactive at any given time. The puddle model predicts that the efficiency falls in proportion to the fraction of reaction centers that have become inactive. In the lake model the efficiency is higher; it does not decrease linearly with the fraction of inactive centers (Fig. 2.6). We shall see that in photosynthetic bacteria the lake model is appropriate. In green plants and algae, at least for the oxygen-evolving Photosystem 2, the behavior suggests an architecture somewhere between the puddle and lake models, with limited energy transfer between neighboring photosynthetic units.

In green plants and algae the definition of a photosynthetic unit is complicated by the fact that there are two kinds of photochemical system. One can ask whether Chl and other pigments act as antenna for Photosystem 1, or Photosystem 2, or both. This question, which will be addressed in Part II of this book, is being resolved by morphological studies of photosynthetic tissues, by the isolation of distinctive pigment–protein complexes from these tissues, and by studies of energy transfer in the intact tissue. At present it appears that there is, in green plants, an antenna pigment–protein complex that serves both photosystems, and also a subantenna for each.

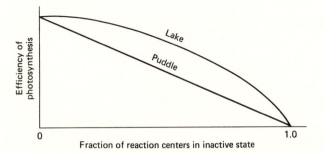

Fig. 2.6. Comparison of photosynthesis efficiency in lake and puddle models, as a function of the fraction of inactive reaction centers.

In green plants there are usually from 200 to 400 molecules of antenna Chl for each reaction center, as in the *Chlorella* cells studied by Emerson and Arnold. In most photosynthetic bacteria the ratio of antenna Bchl to reaction centers is about 50 to 100, but varies with the conditions of growth (see Section 5.4). The green photosynthetic bacteria have an enormous complement of antenna pigment, with about 1000 molecules of Bchl *c* for each reaction center. They seem well adapted to growth in very weak light.

2.3   The evolution of photosynthetic organisms and the impact of oxygen in the biosphere

Speculations on the evolution of photosynthetic organisms begin with the recognition that bacteria are simpler than cells with organized nuclei, and that the photosynthesis of green plants and algae is more advanced (using two distinct photosystems, and evolving O_2 from H_2O) than that of bacteria. Hence it has commonly been proposed that the earliest creatures that bore some similarity to contemporary photosynthetic organisms were bacteria containing Bchl *a*. However, our present knowledge of the biosynthesis of chlorophylls places Bchl *a* several steps beyond Chl *a* in the biosynthetic sequence. In modern photosynthetic bacteria Chl *a* is a precursor of Bchl *a*. The steps from Chl *a* to Bchl *a* are highly selective enzyme-catalyzed reactions; it is unlikely that in earlier times Bchl *a* had evolved on a separate line without going through Chl *a*. Following the principle that the more elaborate biosynthetic pathways reflect successive stages in evolution, J. M. Olson and then D. C. Mauzerall suggested recently that modern photosynthetic creatures have as a common ancestor, about 3.5 billion years ago, a bacterium that contained Chl *a*, probably as an antenna pigment as well as in the reaction centers.

The biosynthetic sequence that supports this idea is outlined sketchily in Fig. 2.7; for a more detailed account see Mauzerall or O. T. G. Jones in the

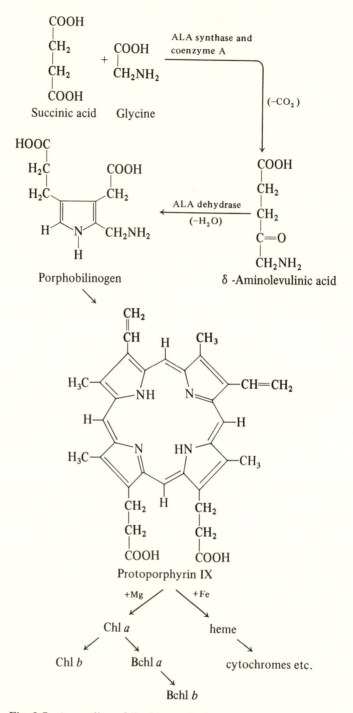

Fig. 2.7. An outline of the biosynthetic pathways of chlorophylls. Insertion of iron instead of magnesium leads to cytochromes and other enzymes involved in photosynthesis and respiration.

suggested reading for this chapter. Succinic acid and glycine are the precursors of δ-aminolevulinic acid in photosynthetic bacteria and also in mammals, but in green plants and algae other precursors including glutamic acid have been implicated. Figure 2.7 omits, among other things, a terminal step in which a long-chain hydrocarbon (phytol, see Fig. 1.1) is attached to complete the formation of Chl *a*, and similar terminal steps that form Bchl *a* and other chlorophylls. The main sequence of Fig. 2.7 is through chlorophyllides, which lack the long hydrocarbon "tail." The steps beyond porphobilinogen, shown by single arrows in Fig. 2.7, are actually multiple steps. Incorporation of magnesium into protoporphyrin IX leads to the chlorophylls; incorporation of iron leads to heme, the prosthetic group of many enzymes including the cytochromes, which are important electron carriers in photosynthesis as well as in respiration.

We start, then, by imagining a primordial Chl *a*-containing bacterium, living in a highly reducing oxygen-free atmosphere. The bacterium could make ATP and convert external nutrients to cell materials, probably by the cyclic mode (Reaction 1.18). Through progressive mutations this creature acquired a noncyclic pattern: the ability to transfer electrons to pyridine nucleotides (or analogous compounds) from more oxidized sources around it. It became able to use ever more oxidized substrates, perhaps progressing through a sequence of nitrogen compounds such as NH_2OH, N_2H_4, NO, NO_2 and NO_2^-, until it gained access to the ultimate substrate, H_2O. But a single photochemical system could not span the gap in redox energies from H_2O to reductants at the level of pyridine nucleotides. The single type of reaction center of the ancestor became differentiated into two that functioned in series, one generating a strong oxidant (ultimately sufficient to take electrons from water, releasing oxygen) and the other retaining the ability to generate a strong reductant, as in Reaction 1.19. In this way a close relative of modern cyanobacteria and other prokaryotic algae had emerged between two and three billion years ago (American billion). The production of O_2 by these microbes had a dramatic effect on the course of evolution on Earth.

Meanwhile another evolutionary branch produced the forerunners of present-day photosynthetic bacteria, containing Bchl and unable to evolve O_2. If the ancestral bacterium contained Chl *a*, it absorbed light strongly in the visible part of the spectrum, with maxima of absorption near 430 and 670 nm and almost no absorption at wavelengths greater than about 720 nm. Imagine a dense population of such creatures in the sea, illuminated from above with diffuse sunlight. The bacteria near the bottom of this swarm would have received very little useful light; the wavelengths best absorbed by Chl *a* would have been screened out by the bacteria nearer the top. A chemical modification of Chl *a*, giving a pigment that absorbs light of wavelengths greater than 720 nm, would have been of great advantage to these bottom dwellers. A few mutational steps of this kind could have led to Bchl *a*, with a maximum of absorption in the near infrared near 800 nm. But in opening a new window in the spectrum so as to escape the problem of mutual shading, these bacteria would have lost the ability to attack highly oxidized substrates and utlimately to take electrons from water. In a suitable environment, oxidized Chl a can be a sufficiently powerful oxidant to attack H_2O, but oxidized Bchl a lacks this potentiality. Thus the evolutionary line that led to Bchl a and other Bchls was a dead end for the development of oxygen-evolving photosynthesis. The shift from Chl a to Bchls may have come early, or it may have arisen as an offshoot of Photosystem 1 (with abandonment of Photosystem 2) in creatures that had already begun to diversify their photosystems.

When photosynthetic microbes first began to evolve oxygen, there was abundant ferrous iron in and around the seas. This fortunate circumstance prevented an immediate rise in the level of O_2 in the biosphere; the highly toxic O_2 was absorbed in a reaction that oxidized Fe^{2+} to Fe^{3+}. The geological evidence of this process is found in banded red strata of Fe_2O_3. All creatures, including those that evolved O_2, had time to erect defenses against the new poison.

Oxygen was toxic because it could inactivate enzymes, especially those that mediate electron transport. And in the light, O_2 in cooperation with a pigment such as Chl can be excited to a highly reactive state that causes the indiscriminate oxidation of many organic compounds. The mechanism of this photo-oxidation involves singlet and triplet states of O_2 and of pigment; the meaning of singlet and triplet states will be described later. For now we can note that a pigment (such as Chl a) enters a singlet excited state when it absorbs light. This is the state in which excitation energy reaches the reaction centers and drives photochemistry. But the excited pigment molecule can also pass from the singlet to a triplet (excited) state, which is longer-lived and more indiscriminately reactive than the excited singlet state. For O_2 the stable ground state is a triplet; it can react with the triplet-state pigment. The products of the reaction are ground-state (singlet) pigment and O_2 in an excited singlet state:

$$P \xrightarrow{h\nu} P^* \longrightarrow P^T$$

then

$$P^T + O_2 \longrightarrow P + O_2^* \tag{2.4}$$

where P and P^* denote the pigment in its ground and excited singlet states, P^T is the triplet state of the pigment, O_2 is the ground (triplet) state, and O_2^* the excited (singlet) state of oxygen. The O_2^* can then attack many organic compounds, including the pigment that promoted the process.

The first line of defense against damage by oxygen was probably provided by carotenoids, the pigments that give carrots and tomatoes their yellow and red colors. Carotenoids are long-chain unsaturated hydrocarbons (a string of

β-Carotene

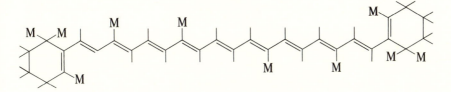

Chromophore of phycocyanin

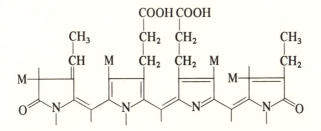

Fig. 2.8. Structures of β-carotene and of the chromophore of phycocyanin, the latter isolated by extracting the chromoprotein with methanol. Carbon atoms are implied at junctions of bonds and H atoms at the ends of unfilled bonds. Methyl, $-CH_3$, is denoted M. In different carotenoids the end groups may or may not be identical, may be closed as rings or open, and may have hydroxy ($-OH$), methoxy ($-OCH_3$), or keto ($=O$) substituents. Also the number of conjugated double bonds can vary. These molecules can be kinked so as to form a variety of isomers.

isoprene residues) with specialized end groups; as an example, the structure of β-carotene is shown in Fig. 2.8. They absorb light of wavelengths below about 550 nm; the long wave limit depends mainly on the number of conjugated double bonds and the isomeric configuration. These pigments occur naturally in all contemporary photosynthetic organisms and in many non-photosynthetic forms of life. Laboratory mutants of algae and photosynthetic bacteria that lack colored carotenoids are killed by the combined presence of light, O_2, and Chl or Bchl.

How do carotenoids give protection against this photo-oxidative damage? First, carotenoids can take energy from triplet-state Chl or Bchl and intercept the reaction (2.4) of the latter with O_2:

$$Chl^T + Carot. \longrightarrow Chl \text{ (grd. state)} + Carot.^T$$

$$Carot.^T \longrightarrow Carot. \text{ (grd. state)} + heat \tag{2.5}$$

Second, carotenoids may react harmlessly with excited (singlet) oxygen, either physically or chemically:

$$\text{Carot.} + O_2^* \longrightarrow \text{Carot.}^T + O_2 \text{ (grd. state)}$$

$$\text{Carot.}^T \longrightarrow \text{Carot. (grd. state)} + \text{heat} \tag{2.6}$$

or:

$$\text{Carot.} + O_2^* \xrightarrow{\text{(harmless chemical cycle)}} \text{Carot. (grd. state)} + O_2 \text{ (grd. state)}$$

$$\tag{2.7}$$

The first of these possibilities seems the simplest and is best documented in vitro and in extracts of photosynthetic tissues.

The very first microbe that evolved O_2 in the light needed some kind of defense against its own O_2. Probably it had already invented a carotenoid as an accessory pigment that could absorb light and transfer the excitation energy to Chl, thus improving its light-harvesting ability in the spectral region 450–550 nm. It may then have been amazed and delighted to find itself immune to its own poisonous O_2. The creatures that lacked carotenoids died out when they tried the experiment of photosynthetic oxygen evolution. And the bacteria that did not evolve O_2 needed carotenoids to survive the onslaught of their O_2-evolving neighbors.

A second defense against the toxicity of O_2 came with the invention of respiration as a way of life. The metabolism of both photosynthetic and non-photosynthetic (fermentative) microbes was modified so that electrons from organic or inorganic substrates could be delivered to O_2 as a terminal acceptor. The O_2 was reduced to H_2O, either directly or by way of H_2O_2. These new organisms could not only dispose of O_2 in their surroundings, they could gain energy and make ATP by this combustion of substrates. In many photosynthetic bacteria the reaction of substrates with O_2 is mediated for the most part by the same electron carriers, quinones and cytochromes, that are used in photosynthetic metabolism.

Some of these microbes abandoned photosynthesis after inventing respiration; some even became precursors of mitochondria, the organelles that carry out respiration and the associated ATP formation in animal and plant cells. Today there are microbes that can perform both photosynthesis and respiration, depending on the availability of light and oxygen. Others can live by either respiration or anaerobic fermentations, and still others remain intolerant of O_2 and must live by photosynthesis and/or fermentation. Green plants and algae are generally capable of respiration, which can help to maintain their store of energy at night. In photosynthetic bacteria that can perform oxidative metabolism (respiration) there is an internal metabolic competition between the two ways of life. The competition favors photosynthesis, so that the respiratory uptake of O_2 is suppressed at moderate to high levels of illumination. Some species of photosynthetic bacteria will not grow in the presence of oxygen because O_2 shuts down the biosynthesis of new Bchl; we shall come back to the regulation of Bchl synthesis in Section 5.4.

These evolutionary processes had a long period in which to get under way, while the ferrous iron in the seas kept the level of oxygen low. About two billion years ago, however, the supply of ferrous iron ran out; nearly all of it had been converted to the ferric form. Oxygen suddenly became more abundant in the seas and in the atmosphere, approaching contemporary levels somewhat less than a billion years ago. Some of this O_2 was converted to ozone by the ultraviolet component of sunlight, forming a diffuse shell of O_3 in the stratosphere. This ozone layer, tenuous as it was (and still is), absorbed most of the sunlight below about 290 nm. This part of the ultraviolet spectrum is the most damaging to life because it is absorbed strongly by nucleic acids and proteins; in fact, the absorption spectrum of ozone matches that of DNA closely, with a maximum near 260 nm.

With ozone providing a screen against the most damaging component of sunlight, life could emerge from the murkier depths and grottoes and begin to flourish near the surface of the sea and even on land. The primitive oxygen-evolving bacteria, precursors of algae and higher plants, may have set out to conquer the world with their new poison, but their weapon backfired. By providing a sunscreen against ultraviolet they paved the way for an explosive diversification and multiplication of all kinds of living things.

2.4   The diversification of photosynthetic pigments

Now let us imagine a primitive photosynthetic bacterium that contained Bchl *a* with an absorption maximum near 800 nm in the near infrared. In dense populations it was not shaded by the creatures that contained Chl *a*, but it was shaded by other members of its own species. One way to evade this shading was to invent a means of swimming preferentially to regions where the light was brighter. Such a capability, called phototaxis, is exhibited by many photosynthetic bacteria today. An example is found in *Rhodospirillum rubrum*, a corkscrew-shaped bacterium with flagella at each end. It can swim equally well in either direction, along its axis. If *Rs. rubrum* swims out of the light into a darker region, it may flip its flagella and reverse its direction, returning to the light. Swimming from darkness into light causes no such response. By this mechanism a spot of light surrounded by darkness acts as a trap for the bacteria. In nature the bacteria that can perform phototaxis will be nearer the light than those that cannot. But this can be a classic evolutionary trap. If every bacterium in a local population has the ability to perform phototaxis, not one is better off than the others. None of them can afford to abandon this ability, and all of them must bear the extra burden of building and maintaining the machinery of phototaxis.

Table 2.1. *Chlorophyllous pigments found in nature*

Pigment	Approximate wavelengths of maximum absorption most commonly found in vivo (nm)[a]
Chl *a*	430, 670–700
Chl *b*	470, 650
Bchl *c*	470, 725–750[b]
Bchl *a*	375, 800–890
Bchl *b*	400, 1020

[a] All these pigments have minor absorption bands in addition to the major ones listed here. Bchls *a* and *b* have a significant band near 600 nm.
[b] There are two or more chemical variants of Bchl *c*.

To continue with our evolutionary fantasy: A bacterium caught in this situation, shaded by the others above it, might have given a figurative shrug of despair – a change in its subcellular architecture. Specifically, the pattern of interactions between Bchl *a* and its surroundings (proteins, lipids, and pigment molecules) may have been altered. This could have caused a change in the energies of ground and excited states of the Bchl *a*, shifting the absorption maximum from 800 nm to a greater wavelength. Pure Bchl *a* dissolved in ether has its long wave absorption maximum at 770 nm. The same Bchl *a*, interacting noncovalently with other molecules of Bchl *a* and with other components in the cell, exists in a variety of physical states in living photosynthetic bacteria. In various species it shows absorption maxima at wavelengths ranging between 800 and 890 nm. Thus a rather trivial modification of physical interactions could open new windows in the solar spectrum, allowing divergent species to coexist without shading each other.

Another evolutionary step was chemical modification, the evolution of a diversity of chemically distinct Chls and Bchls. In contemporary life we recognize chlorophyllous pigments of the kinds listed in Table 2.1. In addition there are more obscure Chls such as Chl *c* and Chl *d*, found in some marine algae.

Aside from the Chls, photosynthetic organisms have evolved two major categories of nonchlorophyllous antenna pigments: carotenoids and phycobilins. Carotenoids are present universally and serve protective as well as light-harvesting functions as discussed earlier. Phycobilins are abundant in blue-green algae and in red algae, where they appear to replace Chl *b* as antenna pigments. The spectral properties of these accessory pigments are given in Table 2.2.

Table 2.2. *Spectral properties of accessory pigments*

Pigment	Approximate wavelength of maximum absorption
Carotenoids	Typically triple-banded spectra in the range 400–550 nm, depending on type and isomer
Phycobilins	
Phycoerythrin	570 nm
Phycocyanin	630 nm
Allophycocyanin	650 nm

The carotenoids are bound noncovalently to the subcellular membranes of photosynthetic organisms, along with the chlorophylls. In the case of phyco-bilins, the chromophores are bound covalently to proteins and form large pigment–protein bodies (phycobilisomes) lying on top of the photosynthetic membranes. These membranes and associated bodies will be described more fully in Part II. The structure of a phycobilin is shown in Fig. 2.8, together with the structure of β-carotene. Phycobilins are open-chain tetrapyrroles; the structure would be approximated by Chl that had lost its magnesium and had been clipped and straightened out.

A glance at Tables 2.1 and 2.2 will show that every part of the visible spectrum, including the near ultraviolet and the near infrared, is absorbed by one antenna pigment or another. Considering the widths of the absorption bands of these pigments, of the order of 50 nm, every niche in the spectrum is utilized by one or more forms of photosynthetic life. The short wave limit of usable light, dictated by the onset of absorption by ozone, is about 300 nm. The long wave limit is set by the fact that water is an effective screen for sunlight at wavelengths beyond about 1150 nm. Water also has an absorption band at 970 nm, and it is interesting to find many photosynthetic bacteria with Bchl *b*, absorbing at 1020 nm in a "window" between the 970 nm and longer wave absorption bands of water.

No higher plant contains Bchl; the leaves in a forest canopy are transparent beyond about 700 nm. This allows them to stay cool while permitting near infrared light and heat to penetrate to the ground, and heat to be dissipated upward through the canopy by radiation.

2.5   Digression: factors that govern the migration of excitation energy among antenna pigments and to photochemical reaction centers

The probability that excitation energy will be transferred from one molecule to another is governed by a rate constant, the reciprocal of an average "transfer time." The probability is high if this transfer rate is large compared with rates of competing processes in the donor molecule, such as fluorescence and radiationless de-excitation (dissipation of the excitation quantum as heat). For two dissimilar molecules the rate of transfer is determined by three factors. First, the rate varies inversely with the sixth power of the separation of the donor and acceptor molecules. This distance is measured between the centers of dipole oscillation in each molecule, for the transitions between ground and excited states. Second, the rate is a function of the cosine of the angle between these dipoles, and of the angle that each dipole makes with a line joining their centers. The rate is maximal for parallel dipoles, zero for mutually perpendicular ones, and two-thirds of the maximum for the average of random orientation. Third, the rate of transfer varies with the extent of overlap between the emission (fluorescence) band of the donor and an absorption band of the acceptor. This third rule is correct if the intensities of fluorescence and absorption are plotted against frequency rather than wavelength.

Because the fluorescence band of a molecule is maximal at a lower frequency than the corresponding absorption maximum (Stokes shift), good overlap is realized if the absorption band of the donor is at a slightly higher frequency than that of the acceptor. This is illustrated in Fig. 2.9 for hypothetical molecules X and Y, with the absorption maximum of Y at higher frequency (greater energy, $E = h\nu$) than that of X. Transfer of excitation energy from Y to X is faster than the reverse, in proportion to the degrees of overlap indicated by the shaded areas. On a wavelength plot the relative positions of these maxima are of course reversed.

Among identical molecules the mechanism of energy transfer depends on the strengths of interactions between the oscillating dipoles. If these interactions are weak compared with the energies of collisions due to thermal motion, the foregoing considerations apply. At the other extreme, if the dipole interactions are strong compared to the spacing of vibrational substates (the energy needed to go from one vibrational substate to another), the rate of migration of energy varies inversely with the third power of the distance between molecules, rather than the sixth power. In that case a "delocalized" description, in which the excited state is the communal property of many molecules, is more appropriate.

In practice, for a solution of randomly oriented pigment molecules such as Chl *a*, the case of weak interaction applies if the concentration is less than about 0.1 M, corresponding to a mean distance between molecules greater than about 2 nm. The case of strong interaction applies if the concentration is greater than about 5 M and the separation is less than about 0.5 nm. The concentrations of chlorophylls and other pigments in photosynthetic membranes fall between these limits, in a region of maximum theoretical complexity. There are pairs or clusters of strongly interacting molecules, with weaker interactions between the clusters, and with the antenna pigments and reaction centers deployed in various ways in different organisms.

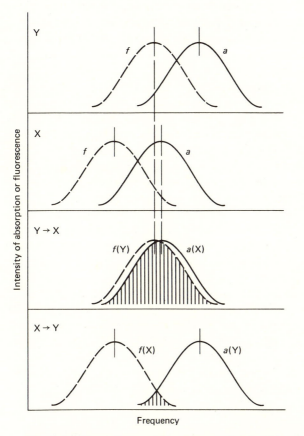

Fig. 2.9. Absorption and fluorescence bands of two hypothetical molecules X and Y. Transfer of excitation energy from Y to X is more rapid than from X to Y, as indicated by the degree of overlap (shaded areas) between the fluorescence band of the donor and the absorption band of the acceptor. These areas are measured as convolution integrals, $\int f(Y) \, a(X) \, dv$ for Y $\longrightarrow$ X. The ordinate is defined quantitatively in Chapter 4.

It is instructive to consider the distribution of excitation energy in a homogeneous set of N molecules of Chl. Anticipating a distinction between antenna and reaction centers, let us label one of these molecules C_0 and the others C, even though for the moment we consider them to be identical. Then with a quantum of excitation executing a random walk among the members of the set, the molecule C_0 will carry the excitation $1/N$ of the time. If the mean lifetime of the excited state (before the quantum is consumed or dissipated in some way) is τ, the molecule C_0 is excited for an average time τ/N. Now let us endow C_0 with a special ability to dispose of the quantum, as in a reaction center where the quantum is used to initiate photochemistry. If a quantum in C_0 can be quenched in this special (photochemical) way in an average time much less than the available time τ/N, the photochemistry is likely to occur. As an example, let the mean lifetime of the quantum in the ensemble, in the absence of special quenching by C_0, be 2×10^{-8} sec. Let $N = 200$, so that any single molecule is excited for an average of τ/N or 10^{-10} sec. Now if the average time for photochemical quenching in C_0 is only 10^{-11} sec, the probability is about 90% that the photochemistry will occur during the available time of 10^{-10} sec. This is best formulated in terms of rates: 10^{11} sec^{-1} for photochemical utilization of a quantum in C_0, and 10^{10} sec^{-1} for loss of this quantum as fluorescence or heat. The photochemical efficiency is then the rate for photochemistry divided by the sum of the rates of all competing processes: $10^{11}/(10^{11} + 10^{10}) = 0.91$ or 91%. Of course the act of photochemical quenching will shorten the mean lifetime of the quantum from τ (the value in the absence of such quenching) to a lesser value, but this way of estimating the probability of photochemical utilization is essentially correct for the simple model we have used.

Now let us consider that the excited state of the special molecule C_0 has an energy different from that of the remainder: The excited state of C_0 has energy E_0, whereas the other ("antenna") C's have excited states of energy E. Intuitively, if $E_0 < E$, C_0 should act as a natural sink for the energy migrating in the system. If E_0 and E are well defined energy levels, and if the system has time to reach thermal equilibrium between these levels, the lower level will be favored over the higher one by a Boltzmann factor,

$$e^{(E-E_0)/kT} \tag{2.8}$$

where E and E_0 are expressed in joules, T is the absolute temperature (degrees Kelvin), and k, the Boltzmann constant, equals 1.38×10^{-23} joule deg^{-1}.

Suppose that C and C_0 have absorption maxima at 680 and 700 nm, respectively, corresponding to the following quantum energies: $E = h\nu = hc/\lambda$; with λ equal to 680 nm or 6.8×10^{-5} cm and with h and c as defined in Section 2.1, $E = 2.91 \times 10^{-19}$ joule. Similarly $E_0 = 2.83 \times 10^{-19}$ joule, and at room temperature (300 K), Formula 2.8 gives a Boltzmann factor of 6.9. The special molecule C_0 has the quantum of excitation not $1/N$ of the time, but about $6.9/N$ of the time. Using the foregoing numerical example with $\tau = 2 \times 10^{-8}$ sec and $N = 200$, the time available for C_0 to initiate photochemistry is about 6.9×10^{-10} sec instead of 10^{-10} sec. If the mean time for photochemical utilization remains 10^{-11} sec, the efficiency of this utilization (computed by a formula based on rates, as before) is 98.6% rather than 91%. As far as photochemical efficiency is concerned, it is as if the antenna were smaller by the factor 6.9 – one reaction center per 29 antenna molecules instead of one per 200.

We shall see later that these numbers were chosen to bear some resemblance to natural systems. In green plant Photosystem 1, about 200 molecules of Chl *a* with maximum absorption near 680 nm serve a reaction center in which the specialized Chl *a* absorbs maximally near 700 nm.

If a photosynthetic membrane contains two or more distinct pools of antenna pigment in contact with each other, similar arguments can be drawn to estimate the proportions of excitation energy in each pool, provided that the lifetimes of the excited states are great enough to allow equilibration among these pools.

2.6   The organization of antenna pigments in relation to energy transfer

The semiquantitative treatment in the foregoing digression was given mainly for pedagogical purposes. It contained simplifying assumptions that are not valid in actual photosynthetic systems. First there was the assumption that the population of excited states can reach thermal equilibrium during the lifetime of excitation. This may be so in the antenna, but in the reaction center the rate of photochemical utilization is too great to afford such equilibration. Once a quantum of excitation has made the jump into a photochemically active reaction center, it will probably be used for photochemistry in 10^{-11} sec or less. The quantum is highly unlikely to remain in the reaction center long enough to reach thermal equilibrium with its surroundings. An active reaction center can be regarded as a nearly irreversible[4] trap for excitation quanta. This trapping process is so efficient that it effectively determines the mean lifetime of an excitation quantum in the antenna, the average time needed for the quantum to encounter and "fall into" the reaction center. For most photosynthetic systems this "transfer time" is of the order of 10^{-10} to 10^{-9} sec. This is long enough to allow fairly complete thermal equilibrium in the antenna, and short enough that losses to fluorescence and heat are kept small ($\lesssim 10\%$) during the journey to the reaction center.

A second simplification in our preliminary treatment was to ignore the distinction between energy transfer among identical antenna molecules and a final jump into the reaction center. One may assume equal sharing of the excitation by each molecule in a homogeneous antenna, but the rate of transfer to the reaction centers from nearby antenna molecules must be computed separately according to the rules for transfer between dissimilar molecules. Not only are the energies of excited states in the reaction center generally different from those in the antenna; the distances between reaction center pigments and the closest antenna molecules are not likely to be the same as the distances between contiguous molecules of antenna pigment. To complicate things further, the reaction centers of known composition contain more than one pigment molecule. For most photosynthetic bacteria, each reaction center contains four molecules of Bchl and two of bacteriopheophytin, a derivative of Bchl that lacks the central magnesium atom. These pigments interact strongly with each other and more weakly with all the neighboring molecules of antenna pigment. A successful computation of energy transfer must consider the whole complex of reaction center pigments as acceptor of the energy, and each neighboring antenna molecule as a potential donor.

Finally, the excited states of these molecules are not described by distinct energy levels such as the E and E_0 of Formula 2.8 (Section 2.5). A single excited electronic state encompasses a continuum of vibrational and rotational substates covering a band of energies (see Fig. 2.2). To estimate the rate of energy transfer into the reaction centers, we should compute the overlap between the fluorescence band of the antenna pigment and all absorption bands of the reaction center (see Fig. 2.9). This procedure replaces the computation of a Boltzmann factor in estimating the rate of transfer to the reaction center. The computation of overlap is not sufficient; we must also know the factors of distance and orientation. Nevertheless, the fluorescence-absorption overlap is useful in comparing the relative advantages of different types of reaction center, in which the excited-state energies bear different relationships to those in the antenna.

Recalling the example of an antenna of Chl a with maximum absorption at 680 nm and a reaction center with an absorption maximum at 700 nm, an analysis of the fluorescence–absorption overlap between antenna and reaction center would show that there is some advantage in this arrangement, when compared with a system in which the excited states of antenna and reaction center have equal energies. For an antenna Chl with maximum absorption at 680 nm, the maximum fluorescence would be near 700 nm. This would overlap well with an absorption band at 700 nm in the reaction center, perhaps two times better than if the absorption band of the reaction center were at 680 nm. Thus the reaction center can act as a "sink" to facilitate energy transfer, but not to the extent of 6.9-fold computed by a simple Boltzmann factor (Formula 2.8) in Section 2.5. Even so, a small increment of photochemical efficiency, say from 91 to 95%, can have great selective force on an evolutionary time scale. Nevertheless there are photosynthetic organisms that appear to throw away this potential advantage. In *Rhodopseudomonas viridis* each reaction center, in which Bchl b has an absorption maximum at 980 nm, is served by an antenna of about 100 molecules of Bchl b absorbing maximally at 1020 nm. The reaction center is a natural "antisink" rather than a sink for excitation energy. The disadvantage of this situation, evaluated from fluorescence–absorption overlap, amounts to a factor of about three. It is as if the organism were obliged to deliver quanta to a single reaction center from an antenna of 300 Bchl b molecules rather than from 100 molecules.

We have seen that we can expect formidable complications when we try to apply simple theories of energy transfer to actual photosynthetic systems. It is not surprising, then, that investigators of photosynthesis rely more on experiment than on theory in exposing the details of energy transfer among antenna pigments and reaction centers. As a general example, consider a sys-

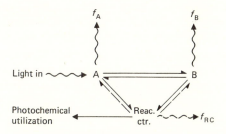

Fig. 2.10. Patterns of energy transfer, fluorescence, and photochemical utilization among antenna components A and B and a reaction center. (See text for discussion.)

tem with two distinct antenna components A and B associated with a reaction center.[5] Each component has its own distinctive spectra of absorption and fluorescence, permitting the selective excitation of any one component and also the identification of fluorescence arising from each. If light is absorbed by antenna component A, the pattern of energy transfer and fluorescence (f_A, etc.) can be sketched (Fig. 2.10). Pathways indicating energy transfer from the reaction center back to the antenna, and fluorescence emitted by the reaction center, have been drawn as dashed arrows in Fig. 2.10 to indicate that these are relatively improbable events.

At any instant the intensity of fluorescence emanating from one component is an indicator of the number of excited molecules in that component because each excited molecule has a certain probability per unit time of emitting a quantum of fluorescence. The fluorescence can therefore show the distribution of excitation energy, either in a steady state (using constant illumination) or as a transient phenomenon after a brief flash of light. The distribution in time can be resolved to a few picoseconds (1 psec = 10^{-12} sec) with contemporary instruments. Using a brief flash absorbed by A, and monitoring f_A and f_B over about 10^{-9} sec, one may be able to see the excitation initially in A, spreading more slowly to B, and then subsiding in both A and B as it is dissipated or trapped by the reaction centers. Fluorescence from the reaction center is usually hard to detect because it is relatively weak and not well resolved spectrally from one or more of the antenna components.

During such measurements, especially with continuous light, the reaction centers are being driven through their photochemical cycle (see Reaction 2.3). At any instant some of the reaction centers are ready to perform photochemistry, whereas others have not yet recovered from a previous cycle of photochemical and "dark" electron transfer. With sufficiently weak light most of the reaction centers are active at any instant, and with sufficiently strong light most of them are driven to an inactive state. If the inactive reaction centers are unable to trap energy from the antenna, the conversion of reaction centers from "active" to "inactive" during constant illumination allows a greater number of excitation quanta to build up in the antenna. This is signaled by a rise in the intensity of fluorescence from the antenna. Thus a set of fluorescence measurements can show the effectiveness of trapping by the reaction centers, the consequences of deleting that trapping, and therefore (in an indirect way) the efficiency of the reaction centers in trapping and utilizing the light energy. One can also measure the fluorescence from preparations in which the reaction centers have been rendered inactive by chemical means; for example, by chemical reduction of the electron acceptor so that it can no longer function as a photochemical electron acceptor.

Such experiments must be interpreted with caution because it cannot be assumed a priori that a reaction center becomes ineffective as a trap for excitation quanta when it enters a state in which it cannot perform "normal" photochemistry. Later we shall examine cases in which photochemically inactive reaction centers retain the ability to trap quanta rapidly and to dissipate the energy as heat. In such cases the antenna fluorescence does not rise to the extent anticipated when the reaction centers are rendered "inactive."[6] A valuable adjunct to experiments of this kind is a genetic mutation in which the reaction centers are missing. Then one can compare the fate of excitation energy in comparable preparations with active reaction centers, with inactive reaction centers, and with no reaction centers at all.

Instead of comparing steady-state levels of fluorescence, one can apply a brief flash of light to systems in which the reaction centers are active, inactive, or absent, and monitor the subsequent decay of fluorescence. If trapping by the reaction centers has been impaired, the fluorescence that signals excitation energy in the antenna will subside more slowly after a flash. This should go hand in hand with a higher level of antenna fluorescence as measured in a steady state.

These experimental uses of fluorometry will be put on a quantitative basis later. For now we can note that the measurements of fluorescence can be complemented by other means. One can measure the efficiency with which quanta of light are used for photochemistry, by determining the rate at which photochemical products are formed and comparing this with the rate at which quanta are being absorbed. By selecting the wavelength of excitation one can compare the efficiency when light is absorbed by one antenna component or another, or directly by reaction centers.

Let us turn now from general methods to a few selected examples of particular findings. In most photosynthetic bacteria that contain Bchl *a* there is an antenna component with a long wave absorption maximum (due to Bchl *a*) at 870–890 nm, close to the long wave maximum of Bchl *a* in the reaction centers. In many species of these bacteria there is a second, distinct antenna component, a pigment–protein complex in which the Bchl *a* shows two absorption maxima near 800 and 850 nm. The former of these antenna components is coupled more closely to the reaction centers than the latter is. Quanta of energy are delivered efficiently to the reaction centers, in a "downhill"

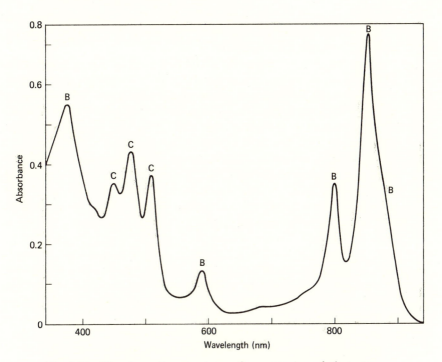

Fig. 2.11. The absorption spectrum of a suspension of photosynthetic membrane fragments ("chromatophores") from the bacterium *Rhodopseudomonas sphaeroides*. Absorption bands due to Bchl *a* are labeled B; those due to carotenoids are labeled C.

cascade through closely overlapping fluorescence and absorption bands: from 800 and 850 nm in one antenna component to 870-890 nm in the other and in the reaction center. Carotenoid pigments in these bacteria deliver excitation energy to Bchl in the antenna with efficiencies ranging from about 15% to 90% in various species.[7] The carotenoids absorb light in the range 400-550 nm. Energy transfer is probably mediated mainly through an absorption band of Bchl *a* near 590 nm, and possibly also through the long wave "tail" of the 375 nm absorption band, extending into the visible spectrum. As an aid in visualizing these spectra, the absorption spectrum of a much-studied photosynthetic bacterium, *Rhodopseudomonas sphaeroides*, is shown in Fig. 2.11. This is a plot of absorbance (see Chapter 4) versus wavelength for a suspension of photosynthetic membrane fragments obtained from these bacteria.

The quantum efficiency of photochemistry, electrons transferred per quantum absorbed, has been measured in reaction centers isolated from *Rp. sphaeroides* and found to be greater than 98%. Corresponding measurements using membrane fragments, with light absorbed by antenna Bchl and transferred to the reaction centers, gave efficiencies of about 90%. The difference reflects energy lost as fluorescence and heat in the antenna, in competition with transfer to the reaction centers.

The green photosynthetic bacteria contain subcellular vesicles full of Bchl *c*, with an absorption maximum at 725 or 750 nm depending on the variety of Bchl *c*. Lying next to these vesicles are packets of Bchl *a* (absorption maximum 810 nm) associated with reaction centers (long wave absorption maximum of the reaction center 840 nm). For each reaction center there are about 80 molecules of Bchl *a* (810 nm) and about 1000 molecules of Bchl *c*. Energy absorbed by this enormous antenna of Bchl *c* is funneled efficiently through the subantenna of Bchl *a* and on to the reaction centers, with good fluorescence-absorption overlap at each stage. These creatures, unexcelled in their large ratio of antenna pigment to reaction centers, are well adapted to growth in weak light. A culture growing in a 30-liter jug, illuminated only by a 25-W tungsten lamp, soon resembles liquid spinach.

In green plants, exemplified by spinach, there is a major antenna component with roughly equal amounts of Chl *a* (absorption maximum near 670 nm) and Chl *b* (650 nm). This light-harvesting Chl *a* + Chl *b* component, termed LH*a*/*b*, serves both Photosystem 1 and Photosystem 2. When LH*a*/*b* absorbs quanta, the partitioning of these quanta between the two photosystems is subject to a regulatory device that provides for balanced operation of the entire system (see Part II). There is evidence that quanta reach the reaction centers of each photosystem by way of separate subantennas containing Chl *a*. In Photosystem 2 the long wave absorption maxima of Chl *a* are at 670–683 nm in the subantenna and 682 nm in the reaction center. The subantenna of Photosystem 1 has Chl *a* absorbing maximally in the range 680–695 nm, and the reaction center has a long wave maximum at 700 nm. Again the "downhill" funneling of excitation energy from antenna pigments to reaction centers is evident, from LH*a*/*b* to each photosystem. Quanta can also migrate from one photosystem to the other, either through LH*a*/*b* or more directly through the subantennas. The amount of this "spillover" is also subject to physiological regulation as described in Part II. About 300 molecules of Chl *a* serve a pair of reaction centers, one of Photosystem 1 and one of Photosystem 2. About 100 molecules of this Chl *a* are in the LH*a*/*b* antenna component; the remainder are in the subantennas.

A particularly nice example of energy cascading "downhill" to reaction centers is found in blue-green algae and red algae.[8] In these algae the accessory phycobilin pigments are packaged in subcellular bodies, the phycobilisomes, which look like tiny knobs attached to the outer surfaces of the photosynthetic membranes. These bodies contain phycoerythrin, phycocyanin and allophycocyanin in a layered arrangement, with phycoerythrin (absorption maximum 570 nm) outermost, phycocyanin (630 nm) next, then allophycocyanin (650 nm), followed by allophycocyanin B (670 nm) closest to Chl *a* (670–680 nm) in the membrane. This Chl *a* acts as an antenna for the reaction centers of Photosystems 1 and 2. Light absorbed by the phycobilins is transferred (through Chl *a*) preferentially to Photosystem 2, but energy can be transferred readily between the two photosystems.

These algae, in the form of multicellular filaments, afforded an elegant demonstration of energy transfer in one of many imaginative experiments made by T. W. Engelmann in the latter part of the nineteenth century. This was long before the transfer of excitation energy was first described as a physical phenomenon (by G. Cario and J. Franck in 1923, using mixed vapors of mercury and thallium) or implicated explicitly in photosynthesis (by H. J. Dutton and W. M. Manning in 1941). Engelmann projected a microspectrum, viewed through a microscope, along a single algal filament. He expected that in those regions where the wavelength was most favorable for photosynthesis, the oxygen concentration would be greatest. To test this he introduced a population of aerotactic (oxygen-seeking) bacteria around the filament and let the preparation rest in the dark until it had become free of oxygen as a result of respiration by the bacteria. Then when the illumination (in the form of a spectrum) was turned on, the bacteria began to congregate, by their movement toward oxygen, around those parts of the filament that were evolving oxygen the most rapidly. Engelmann thus found that light in the region 570-630 nm, absorbed mainly by phycobilins, was particularly effective for photosynthesis. This provided the first evidence that accessory pigments, other than chlorophylls, could function in photosynthesis.

Details of the subcellular structures that effect these patterns of energy transfer in various photosynthetic organisms will comprise the main subject of Part II.

SUGGESTED READINGS

Clayton, R. K. (1971). *Light and Living Matter: A Guide to the Study of Photobiology*, Vol. 1 (The Physical Part), Chapter 2, pp. 5-62. McGraw-Hill, New York. Reprinted in 1977 by R. E. Krieger, Huntington, New York.

Stoeckenius, W. (1978). Bacteriorhodopsin. In *The Photosynthetic Bacteria*, R. K. Clayton and W. R. Sistrom eds., Chapter 29, pp. 371-594. Plenum, New York.

Emerson, R., and Arnold, W. (1932). A separation of the reactions in photosynthesis by means of intermittent light. *J. Gen. Physiol. 15*, 391-420.

Emerson, R., and Arnold, W. (1932). The photochemical reaction in photosynthesis. *J. Gen. Physiol. 16*, 191-205.

Clayton, R. K. (1971). *Light and Living Matter*, Vol. 2 (The Biological Part), Chapter 1, pp. 1-66. McGraw-Hill, New York. Reprinted as above.

Jones, O. T. G. (1978). Biosynthesis of porphyrins, hemes, and chlorophylls. In *The Photosynthetic Bacteria*, R. K. Clayton and W. R. Sistrom, eds., Chapter 40, pp. 751-78. Plenum, New York.

Olson, J. M. (1970). The evolution of photosynthesis. *Science 168*, 438-46.

Mauzerall, D. C. (1978). Bacteriochlorophyll and photosynthetic evolution. In *The Photosynthetic Bacteria*, R. K. Clayton and W. R. Sistrom, eds., Chapter 11, pp. 223-32). Plenum, New York.

Cloud, P. (1974). Evolution of ecosystems. *Am. Sci. 62*, 54-66.

Schopf, J. W. (1978). The evolution of the earliest cells. *Sci. Amer. 239*, 110-38.

Clayton, R. K. (1971). *Light and Living Matter*, Vol. 2, pp. 203-14 (in Chapter 5). McGraw-Hill, New York. Reprinted as above.

Sistrom, W. R., Griffiths, M., and Stanier, R. Y. (1956). The biology of a photosynthetic bacterium which lacks colored carotenoids. *J. Cell. Comp. Physiol. 48*, 473-515.

Zechmeister, L. (1962). *Cis-Trans Isomeric Carotenoids, Vitamins A and Arylpolyenes*. Springer-Verlag, Vienna.

Liaaen-Jensen, S. (1978). Chemistry of carotenoid pigments. In *The Photosynthetic Bacteria*, R. K. Clayton and W. R. Sistrom, eds., Chapter 12, pp. 233-47. Plenum, New York.

Förster, T. (1948). Zwischenmolekulare Energiewanderung und Fluoreszenz. *Ann. Physik 2*, 55-75.

Paillotin, G. (1972). Transport and capture of electronic excitation energy in the photosynthetic apparatus. *J. Theor. Biol. 36*, 223-35.

Duysens, L. N. M. (1952). Transfer of excitation energy in photosynthesis. Doctoral thesis, State University, Utrecht, The Netherlands.

Zankel, K. L., and Clayton, R. K. (1969). "Uphill" energy transfer in a photosynthetic bacterium. *Photochem. Photobiol. 9*, 7-15.

Thornber, J. P., Trosper, T. L., and Strouse, C. E. (1978). Bacteriochlorophyll in vivo: Relationship of spectral forms to specific membrane components. In *The Photosynthetic Bacteria*, R. K. Clayton and W. R. Sistrom, eds., Chapter 7, pp. 133-60. Plenum, New York.

Engelmann, T. W. (1881). Neue Methode zur Untersuchung der Sauerstoff Ausscheidung pflanzlicher und tierischer Organismen. *Bot. Zeitung 39*, 441-8.

Engelmann, T. W. (1884). Untersuchungen über die quantitativen Beziehungen zwischen Absorption des Lichtes und Assimilation in Pflanzenzellen. *Bot. Zeitung 42*, 81-93, 97-105.

3 The cooperation of two quanta or two photosystems in photosynthesis

3.1 The quantum efficiency of photosynthesis: a renowned former controversy

The question of the quantum efficiency of photosynthesis took on specific meaning with van Niel's formulation of the photochemical process (see Reaction 1.6). The assimilation of one molecule of CO_2 coupled with the evolution of one molecule of O_2 requires, as "primary" photoproducts, four strong reducing equivalents separated from four strong oxidizing equivalents. How many quanta are needed to effect this? If the answer is four, then one quantum can bring about the separation of one strong reducing equivalent and one strong oxidizing equivalent. If eight quanta are needed per O_2 and per CO_2, then two quanta must cooperate somehow in this separation of oxidant and reductant.

After 40 years this question seems to be resolved, but only after having drawn many scientists into a prolonged and bitter controversy. O. Warburg and his collaborators, measuring photosynthesis in suspensions of the green alga *Chlorella*, reported that four quanta suffice for the evolution of one molecule of O_2. Warburg used the manometric technique that he had helped to develop, measuring the O_2 evolved by algae in a closed vessel through the attendant increase of pressure. Uptake of CO_2 was compensated by a buffer solution of bicarbonate that maintained a constant partial pressure of this gas.

Warburg held to his claim of 4 quanta per O_2 while many other investigators were reporting values in the range of 6 to 12 quanta/O_2, using a variety of techniques including Warburg's manometry. The controversy reached its height during the 1940s and was sustained for many years, with a flood of measurements punctuated by polemics involving Warburg and his adherents on one side and the majority of the "photosynthesis community" (notably R. Emerson, J. Franck, H. Gaffron, and E. Rabinowitch) on the other. Warburg made the legitimate claim that the highest measured efficiencies were the most reliable and the most pertinent to the mechanism of photosynthesis. He asserted that everyone else was somehow mistreating the algae and observing lower efficiencies (higher quantum requirements). Emerson and others showed that by using Warburg's methods they could obtain data similar to

Warburg's; it was a question (to which we shall return) of how the raw data should be interpreted.

Franck argued that Warburg's claim was on weak ground in terms of energetics. A quantum at 680 nm, near the long wave absorption maximum of Chl a, has an energy of 2.9×10^{-19} joule; the combined energy of four such quanta is 11.6×10^{-19} joule. The assimilation of one molecule of CO_2 as sugar, accompanied by the evolution of one molecule of O_2 from H_2O, represents a storage of free energy equal to 7.8×10^{-19} joule (equivalent to 168 kcal mole^{-1}). Warburg's claim could be restated: The fraction of quantum energy finally stored as chemical free energy is 7.8/11.6 or 67%. This is barely within the limits of conceivable energy efficiency for the primary photochemistry, whether interpreted mechanistically (some energy must be given up in order to stabilize the photoproducts against wasteful recombination) or in terms of adherence to the second law of thermodynamics. All the secondary chemical processes must then operate without losses, like a frictionless machine, if the loss of energy is accounted for fully at the "primary" level. This suggested to Warburg a kind of perfection in nature that appealed to him, so much so that he eventually welcomed "new and improved" data purporting to show that only 2.7 quanta are needed for each molecule of O_2 evolved, thereby eliminating all losses of energy. At that point Warburg's position was just exactly within the bounds of the conservation of total energy (first law of thermodynamics) and clearly in violation of sensible ideas of mechanism. He was forced to reject van Niel's hypothesis and return to the older speculations of Willstätter in which CO_2 and H_2O are rearranged while in intimate contact with illuminated Chl.

We now have general agreement that 8 to 10 quanta are needed for the photosynthetic evolution of one molecule of O_2 and assimilation of one CO_2. The claims and arguments of Warburg have subsided, but only after much effort had been expended by many scientists, especially Emerson, on this one question. A comprehensive and excellent account of this matter has been written by Gaffron (see the suggested reading for this chapter). But how did the disagreement arise?

In Warburg's manometric measurements the suspensions of algae were exposed to alternating periods of light and darkness. In the dark the algae consumed O_2 by respiration, but in the light the photosynthetic evolution of O_2 exceeded the respiratory uptake. The net evolution of O_2 due to photosynthesis was obtained by correcting for respiratory uptake as measured in the dark. Warburg assumed that the rate of respiration was not affected by light, and this may have been his first mistake. New techniques[1] that distinguish between photosynthetic evolution of O_2 and respiratory uptake indicate that light can inhibit respiration, especially at the average intensity to which the algae were exposed in Warburg's measurements. By applying too large a correction for respiration, Warburg may have overestimated the rate of photosynthetic oxygen evolution.

A more obvious and serious source of disagreement lay in the fact that the rate of O_2 evolution was not constant during the period of illumination, nor was the rate of O_2 consumption constant during the dark period in Warburg's measurements. Each light period began with a gush of O_2 evolution, about twice the steady rate attained after the first minute or so. Each dark period began with a "gulp" in which O_2 was consumed at a rate higher than the subsequent steady rate in the dark. Warburg computed the quantum requirement on the basis of the initial rate of O_2 evolution in the light, during the gush. Emerson, Gaffron, and others used the steady rate attained later. They argued that the O_2 gush reflected an abnormal situation in which the evolution of O_2 was coupled to the utilization of metabolic reserves stored previously in the dark.[2] This difference of interpretation was enough to account for the disagreement between Warburg and the others.

By the 1950s it was generally accepted that eight (or more) quanta are needed for the evolution of one O_2 molecule and the concomitant assimilation of one CO_2. In the framework of van Niel's hypothesis, the question became "How do two quanta cooperate in the separation of one strong reducing equivalent and one strong oxidizing equivalent?"

3.2 Evidence for two distinct components of chlorophyll and two photochemical systems in oxygen-evolving photosynthesis

In the 1950s, while scientists wondered how two quanta cooperate to produce the primary photochemical products in photosynthesis, some curious facts began to emerge about the roles of phycobilins and Chl in blue-green algae. L. R. Blinks, F. T. Haxo, L. N. M. Duysens, and others were determining action spectra (measurements of the relative effectiveness of different wavelengths of light; see Chapter 4) for Chl fluorescence and for photosynthesis. They obtained the surprising result that light absorbed by phycobilins is more effective in promoting Chl fluorescence than is light absorbed by Chl itself. This could only mean that the algae contain two components of Chl, one with a higher yield of fluorescence than the other. Light energy absorbed by the phycobilins is transferred preferentially to the more highly fluorescent component; light absorbed by Chl is partitioned between the two components. Action spectra for photosynthesis also showed that light absorbed by phycobilins is more effective than light absorbed directly by Chl.[3] Thus the more strongly fluorescent component of Chl was identified also as the more active for photosynthesis. With the two Chl components denoted Chl_1 and Chl_2, the data fit a model of this form:

$$Chl_1 \text{ (low fluorescence; inactive for photosynthesis)}$$

$$Light \rightsquigarrow Chl_2 \text{ (high fluorescence; active)}$$

$$Phycobilins$$

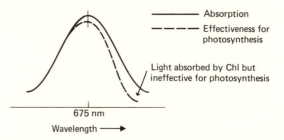

Fig. 3.1. Plot showing photosynthetic ineffectiveness of light absorbed in the long wave edge of the long wave absorption band of Chl.

Another curious finding, discovered by Emerson and soon confirmed for a variety of algae and plants, was the "red drop" effect. Action spectra showed that light absorbed in the long wave edge of the long wave absorption band of Chl is peculiarly ineffective for photosynthetic oxygen evolution (Fig. 3.1). Thus the "inactive, weakly fluorescent" component of Chl implicated in the studies with blue-green algae could be equated with a "long wave" form that is predominant in the absorption spectrum at wavelengths greater than about 680 nm. The two forms of Chl do not differ enough to be resolved as separate peaks in an absorption spectrum.

The difference between the two forms of Chl is not merely a quantitative difference in photosynthetic efficiency. Even with saturating illumination, providing the greatest possible rate of photosynthesis, the rate was found to decline at wavelengths greater than about 680 nm but well within the absorption band of Chl.

A third curiosity emerged in the "chromatic transients" discovered by L. R. Blinks. Measuring oxygen evolution by algae, Blinks was attempting to adjust the intensities of light of two different colors (e.g., red and green) to produce equal rates of photosynthesis, so that the substitution of one color for the other would cause no change in the rate of oxygen evolution. Even if the intensities were adjusted to give equal rates in the steady state, there was always a transient fluctuation following the substitution. Green light after red caused a brief gush of O_2 evolution and then a transitory subnormal phase before a constant rate was attained. Red after green caused a temporary decline of the rate prior to the steady state. One might say that these algae are not color-blind; different wavelengths have qualitatively different effects that cannot be neutralized by adjusting the intensities.

The most dramatic of these curious effects was revealed in Emerson's discovery of the phenomenon of enhancement. Using the favorite research material of that time, the green alga *Chlorella*, Emerson tested the effects of light of two wavelengths given separately and then together. In some cases there

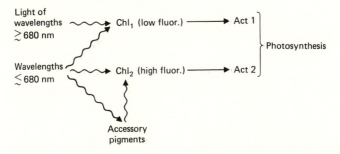

Fig. 3.2. A model to explain the qualitatively different effects of light of different wavelengths in promoting photosynthesis, as perceived in the 1950s.

seemed to be a synergistic action; the rate of O_2 evolution with both qualities of light superimposed was greater than the sum of the rates produced by each light acting alone. The synergism was especially pronounced if one of the two wavelengths was greater than about 680 nm, in the region of the "red drop" effect. More detailed measurements indicated that the poor effectiveness of "far red" light (>680 nm) was corrected by mixing in some shorter wave light.

All of these findings could be accommodated by a model that invoked two distinct photochemical acts, mediated by the two components of Chl, with both needed for complete photosynthesis. The model is outlined in Fig. 3.2. Far red light is absorbed only by Chl_1, producing only "photochemical act 1." Shorter wave light, absorbed by both components of Chl and by accessory pigments (phycobilins in blue-green algae; Chl *b* in other algae and plants), can drive both photochemical acts, and preferentially "act 2." The enhancement phenomenon arises because neither photoact is sufficient by itself. How a balance between the two photoproducts is achieved in nature is a question to which we shall return in Part II.

At the time this was little more than phenomenology, but some properties of the hypothetical products of act 1 and act 2 could be defined. J. Myers and C. S. French observed enhancement even when the two qualities of light were separated in time. Flashes of far red and shorter-wave light could produce enhancement in *Chlorella* if the two flashes were given in quick-enough succession. The limiting dark interval was about 6 sec for a far red flash after a green flash, and about 1 min for green after far red. It appeared that the product of act 2 had a lifetime of about 6 sec; during this time it could interact fruitfully with the product of act 1. The product of act 1 persisted as long as a minute.

The discovery of this "sequential enhancement" laid to rest a variety of theories invoking two-quantum cooperation at a physical level. As an exam-

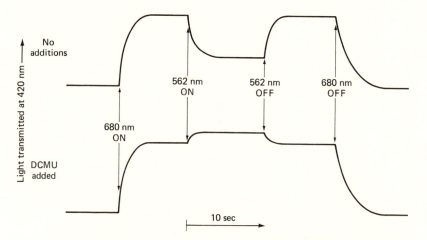

Fig. 3.3. An experiment reported by L. N. M. Duysens and collaborators, in which a suspension of *Porphyridium* (an alga containing phycoerythrin as the major accessory pigment) was monitored with a weak beam of 420 nm light. An increase of the light transmitted by the suspension signaled the oxidation of a cytochrome in the algae. Strong light at 680 and/or 562 nm was applied to the algae during the measurement. The 680 nm light caused oxidation of the cytochrome; superposition of 562 nm light caused a partial re-reduction. The latter effect was abolished by the herbicide 3-(3,4-dichlorophenyl)-1,1 dimethylurea (DCMU).

ple, Franck had proposed that one quantum promotes Chl into a triplet state, and a second quantum, delivered to the same Chl, drives it into a state of still greater energy. The state of higher energy is required for photosynthesis. But these physical states of Chl are far too transitory to qualify as products of "act 1" and "act 2"; their lifetimes are far shorter than 1 sec, let alone 6 or 60.

By 1960 the stage was set for a new and apparently correct interpretation of the two distinct photochemical acts that cooperate in the photosynthesis of green plants and algae. The new interpretation, the "Z scheme" outlined in Reaction 1.19, was consistent with a requirement of 8 to 10 quanta for the evolution of one O_2 and showed the nature of the "two-quantum cooperation" implied by such a quantum requirement. Van Niel's "strong reductant" (Reaction 1.6) is the reducing entity produced by Photosystem 1, and the "strong oxidant" is the oxidizing product of Photosystem 2.

The Z scheme was first suggested by R. Hill and F. Bendall; the most compelling experimental support came from experiments reported by Duysens in 1961. The decade of the 1950s had brought great refinement to the technique of sensitive differential absorption spectrometry. The light-induced alteration of molecules, present in trace amounts in photosynthetic tissues,

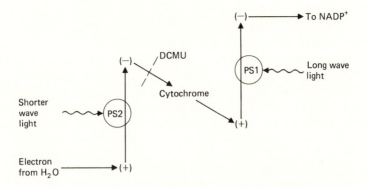

Fig. 3.4. A model for the cooperation of Photosystem 1 and Photosystem 2 in oxygen-evolving photosynthesis, explaining the experimental results shown in Fig. 3.3. DCMU blocks the flow of electrons from Photosystem 2 to the cytochrome. This model is equivalent to the Z scheme proposed by R. Hill and F. Bendall in 1960.

could in some cases be detected by changes of their absorption spectra. As an example, the oxidation and reduction of cytochromes, secondary electron carriers in photosynthesis, could be detected by changes in the absorption of light in the regions 400–430 nm and 550–560 nm (see Fig. 4.7, Section 4.4).

Duysens used this technique to measure the light-induced oxidation and reduction of a cytochrome in a suspension of *Porphyridium* algae. While measuring the light transmitted by the sample at 420 nm, using a weak monochromatic measuring beam, he superimposed (at right angles to the measuring beam) a beam of strong light to elicit photochemistry. This exciting light was of wavelength either 680 nm or 562 nm, or both together. Oxidation of the cytochrome resulted in an increase of the 420 nm light transmitted by the sample, and reduction a decrease. Duysens's principal findings are shown in an idealized form ("noise" deleted) in Fig. 3.3 Long wave (680 nm) light caused the oxidation of the cytochrome. Superposition of shorter wave (562 nm) light caused a partial re-reduction. Thus the two qualities of light had opposite effects on the redox poise of the cytochrome. The herbicide DCMU (3-(3,4-dichlorophenyl)-1,1-dimethylurea), which abolished photosynthetic oxygen evolution, eliminated the reducing action of shorter wave light, so that the effect of this light became like that of long wave light, but less pronounced.

Duysens's interpretation, consistent with the model proposed by Hill and Bendall, is shown in Fig. 3.4. Far red light removes electrons from the cytochrome; these electrons go to the primary oxidant made by Photosystem 1. Shorter wave light delivers electrons to the cytochrome from the reducing side of Photosystem 2. DCMU blocks the flow of electrons from Photosystem 2 to the cytochrome. This interruption of electron transport accounts for the inhibition of oxygen evolution.

Evidence supporting this Z scheme was soon found in a variety of algae and green plant tissues, especially spinach chloroplasts. A detailed elaboration of the events outlined in Fig. 3.4 will be the main subject of Part III.

The Z scheme is superficially consistent with a requirement of eight quanta per O_2 and CO_2. Four quanta captured by each photosystem are needed in order to drive four electrons from H_2O to NADP.

SUGGESTED READINGS

Gaffron, H. (1960). Energy storage: Photosynthesis. In *Plant Physiology*, F. C. Steward, ed., Vol. 1B, pp. 3–277. Academic Press, New York. [This six-volume treatise should not be confused with the current periodical *Plant Physiology.*]

Blinks, L. R. (1954). The photosynthetic function of pigments other than chlorophyll. *Annu. Rev. Plant Physiol. 5*, 93–114.

Emerson, R., and Lewis, C. M. (1943). The dependence of the quantum yield of *Chlorella* photosynthesis on wavelength of light. *Am. J. Bot. 30*, 165–78.

Blinks, L. R. (1957). Chromatic transients in photosynthesis of red algae. In *Research in Photosynthesis*, H. Gaffron et al., eds., pp. 444–9. Interscience, New York.

Emerson, R. (1958). The quantum yield of photosynthesis. *Annu. Rev. Plant Physiol. 9*, 1–24.

Franck, J. (1957). A theory of the photochemical part of photosynthesis. In *Research in Photosynthesis*, H. Gaffron et al., eds., pp. 142–6. Interscience, New York.

Hill, R., and Bendall, F. (1960). Function of two cytochrome components in chloroplasts: A working hypothesis. *Nature (London) 186*, 136–7.

Duysens, L. N. M., Amesz, J., and Kamp, B. M. (1961). Two photochemical systems in photosynthesis. *Nature (London) 190*, 510–11.

4 Major digression: molecular physics and spectroscopy; quantum energy and redox energy; measurements involving light

4.1 Covalent bonding; electron spin, radicals, and triplet states; microwave spectroscopy

The states of molecules can be described in terms of the states of the component electrons and nuclei. The detailed quantum state of an electron in a molecule includes a specification of its energy and various aspects of its angular momentum: the magnitude and direction[1] of angular momentum associated with orbital motion, and the direction of angular momentum associated with the electron's spin about its axis (relative to directional aspects of other parts of the molecule, such as the spins of other electrons).

The common saying that two things cannot be in the same place at the same time has its counterpart, in quantum theory, in the Pauli Exclusion Principle: No two particles can be in the same detailed quantum state. If two electrons in a molecule have their spins in opposite directions, they can be in the same state in every other respect. The electrons in atoms and molecules tend to occupy the states of least energy, filling these low-lying levels with pairs of electrons having opposed spins. In an atom or a radical there may be an odd electron left over, but in a covalently bonded organic molecule that is not a radical, and is not excited, there are no unpaired spins. The total angular momentum due to spin is then zero because in each pair the opposed spins, added as vectors, cancel each other.

If a single electron is removed from or added to an organic molecule, the molecule acquires one "unpaired" electron and is referred to as a radical. The spin of the unpaired electron is assigned the formal value $\frac{1}{2}$ in the mathematics of quantum mechanics. An alkali metal also has spin $= \frac{1}{2}$ because of the unpaired valence electron.

If a molecule is excited, so that one electron is raised to a state of higher energy, that electron is no longer constrained by the Pauli Exclusion Principle to have its spin opposite to that of its former partner. The rules of quantum mechanics specify that the spin of the promoted electron may be either antiparallel or parallel to that of its former partner. In the antiparallel case the total spin is zero, as in the ground state. In the parallel case the sum of the two parallel electron spins is $\frac{1}{2} + \frac{1}{2}$ or unity, in the formalism of quantum mechanics.[2]

The spin of an electron is measured by its magnetism; a spinning electron constitutes a circulating electric charge and generates a magnetic field. The electron interacts with other sources of magnetism in its surroundings, being attracted or repelled by an external magnetic field, depending on the alignment of the electron spin relative to the field. Consequently a molecule with net spin can acquire greater or lesser energy when placed between the poles of a magnet. The energy is lowered slightly when the spin is parallel to the field of the magnet, raised when it is antiparallel, and unaltered when it is perpendicular. A single energy state can thus be split into two or more by the magnet.

The rules of quantum mechanics provide that when the spin equals $\frac{1}{2}$, the system (an organic radical, for example) can interact with the magnetic field in just two ways: parallel or antiparallel. A single state is therefore split into two when the magnet is switched on, and this case (spin = $\frac{1}{2}$) is accordingly called a doublet. When the spin equals unity, as in an excited molecule with two parallel electron spins, three kinds of interaction are allowed: parallel, perpendicular, and antiparallel. A single state is split into three by the magnet, and this kind of state (spin = 1) is called a triplet. When the spin is zero, as with most organic molecules in their ground states, the magnetic field has no effect, and the state is called singlet.

The splitting of energy states in a magnetic field can be detected by causing and observing transitions between these newly formed substates. Just as a quantum of visible light can promote major electronic transitions in a pigment, so can a quantum of suitable energy raise a molecule (radical or triplet) from the lower to the upper of two substates generated by splitting in a magnetic field. The energy gap is proportional to the strength of the magnetic field; with magnets of convenient laboratory size the energy gap is of the order of 10^{-23} joule, corresponding to "microwave" quanta: frequency about 10^{10} sec^{-1}, wavelength a few centimeters. In applying this technique it is convenient to maintain a fixed frequency of microwave radiation and vary the magnetic field strength, observing the points at which strong absorption of quanta occurs and thereby identifying the energy gaps between the magnetic substates. Microwave spectroscopy differs in this respect from optical absorption spectroscopy, wherein one observes changes in absorption as the frequency (or wavelength) is varied.

The spin magnetic moment of an electron equals $\frac{1}{2}g\beta$, where β (the Bohr magneton) equals $eh/4\pi mc$ or about 10^{-27} joule/gauss, and g is a numerical constant, 2.00232 for a free electron. For a single unpaired electron ($s = \frac{1}{2}$) the energy gap between the magnetically split doublet states is $\Delta E = g\beta H$, where H is the magnetic field strength. In a molecule, the spin of an unpaired electron interacts not only with an external magnetic field but also with magnetic forces due to other electrons and nuclei in the micro-environment. As a result of these interactions the single g value of 2.00232 (free electron) is replaced by an assortment of g values, usually near 2.0. A microwave absorption spectrum can therefore show an abundance of detail (band broadening and multiple bands) that helps to identify the molecule and its state.

For technical reasons, most microwave spectrometers display the first derivative of the absorption spectrum (da/dH vs. H) rather than the absorption spectrum itself (a vs. H). The centers of absorption bands appear as zero-crossings ($da/dH = 0$) in these derivative spectra. The significant parameter for interpretation is the g value that relates ΔE to H; the abscissa of the spectrum can be calibrated in terms of g by recording the spectra of radicals whose g values (at the absorption maxima) are known.

Microwave spectroscopy applied in this way to the electron spins of atoms and molecules is called electron spin resonance (ESR) spectroscopy, with which term electron paramagnetic resonance (EPR) is synonymous. When applied to the spins of nuclei the technique is called nuclear magnetic resonance (NMR) spectroscopy. In the special technique of electron–nuclear double resonance (ENDOR), the interactions between unpaired electrons and nuclei are examined selectively. This gives more specific information about the nuclei that lie within the spatial distribution (expressed by the square of the wave function) of an unpaired electron.

When a molecule is excited from its electronic ground state, it is most likely to enter a singlet excited state, in which the spin of the promoted electron remains antiparallel to that of its former partner, and the total spin remains zero. A triplet state is not likely to be formed during excitation because the spin of the promoted electron must become reversed from its former alignment. This event requires some sort of magnetic interaction with the surroundings; it can be facilitated, for example, by the magnetic field of the nucleus of a heavy atom. The intense absorption and emission of 254 nm light by mercury is due to transitions between a singlet and a triplet state.

Passage into a triplet state is more likely to occur from an excited singlet state than from a ground state because the electron distribution is more widespread, increasing the interaction with nearby magnetic fields and decreasing the magnetic coupling by which a pair of electrons tends to retain antiparallel spins. If Chl a in ether solution is excited, there is roughly a 33% chance that the molecule will emit fluorescence and return to the ground state, and 67% that it will go over into a triplet state. Eventually the triplet-state Chl will return to the ground state.

Once in a triplet state, a molecule is especially reactive for two reasons. First, the lifetime of this excited state is far greater than that of the corresponding excited singlet state, often about 10^{-3} sec rather than 10^{-8} sec, because the return from triplet to ground, like the direct excitation from ground to triplet, is a relatively improbable event, requiring a reversal of spin orientation. Second, in a triplet state the electron distribution is more widespread than in the corresponding singlet state. A molecule in a triplet state

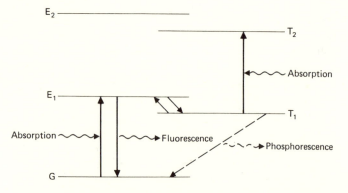

Fig. 4.1. An energy diagram showing singlet and triplet states in a molecule, simplified by omitting vibrational and rotational substates and by not showing all possible transitions. G, E_1, and E_2 are ground and excited singlet states; T_1 and T_2 are triplet states corresponding to E_1 and E_2. Highly probable "dipole-allowed" radiative transitions are shown by wavy arrows. Transitions between E_1 and T_1 are usually less probable, and that from T_1 to G is even less so. The return from T_1 to G may be radiative, with a quantum of phosphorescence emitted. This phosphorescence can be distinguished from the fluorescence emitted in $E_1 \longrightarrow G$ by its greater wavelength, corresponding to a smaller energy gap. A delayed fluorescence can arise if a molecule in the state T_1 returns to E_1 and then to G.

is therefore more likely to exchange an electron with its surroundings. Nevertheless, as we shall see in Part III, the photochemistry of photosynthesis is mediated by the excited singlet state of Chl and not by the triplet state. This can be rationalized by comparing rates of energy transfer and photochemical utilization (all involving the excited singlet state of Chl) with the much slower rate of entry into the triplet state. The yield of triplet Chl upon excitation in vitro is comparatively high because in the absence of photochemical utilization, the excited singlet state has a much greater lifetime. In the living tissue, photochemistry competes effectively against triplet formation as well as against fluorescence.

Triplet states of molecules can be detected optically as well as by ESR. For each excited singlet state there is a corresponding triplet state of lower energy, as sketched in Fig. 4.1. The transition from T_1 to T_2 does not require a change of spin orientation and is therefore in the category of highly probable radiative transitions. A population of molecules in the state T_1 will show a characteristic absorption band corresponding to $T_1 \longrightarrow T_2$, and other bands for transitions from T_1 to still higher triplet states. The return from T_1 to ground may also be attended by a weak phosphorescence, which is distinguished from the fluorescence ($E_1 \longrightarrow G$) by its greater wavelength and lifetime. Lifetime alone is not a sufficient criterion, because a delayed fluorescence can arise if a molecule is promoted from T_1 back to E_1 and thence to ground.

Finally, triplet states can be characterized by a mixture of optical and microwave spectroscopy called optically detected magnetic resonance (ODMR). The triplet state is split into substates by an external magnetic field. The rate of interconversion between T_1 and E_1 is different for each of these magnetically induced substates, and the relative number of molecules in each substate can be manipulated by applying microwave radiation. As a result, the intensity of fluorescence ($E_1 \longrightarrow G$) can be modulated by the combination of external magnetic field and microwave radiation. This kind of experiment can give more information than either optical or microwave spectroscopy alone.

4.2 Molecular spectroscopy: electron orbitals, energy states, and optical absorption spectra

To a first approximation, the chemical and optical properties of atoms are determined by the wave functions (counterparts of the classical orbits) of the outer or valence electrons. The electronic states of molecules can be described similarly in terms of "molecular orbitals." These orbitals are classified in terms of symmetry. For a hydrogen atom in its ground state the orbital of the single electron has spherical symmetry; in the notation of spectroscopists this is an *s* state.[3]

In a diatomic molecule an orbital embracing both atoms with axial symmetry is classified σ, the molecular counterpart of *s*. A simple example, that of the molecular ion H_2^+, is illustrated in Fig. 4.2. For an isolated atom of hydrogen it is immaterial whether the wave function ψ is assigned a positive or a negative value; the spatial distribution is determined by ψ^2. The sign of ψ becomes important when atomic orbitals fuse to become molecular ones. In the case of H_2^+, equal signs give a bonding orbital and opposite signs an antibonding orbital as sketched in the figure. The bonding orbital is the more stable; it corresponds to the ground state of the H_2^+ ion. In the less stable antibonding case there is a node, a plane between the two nuclei where the electron has zero probability of existing. In this state the ion is more likely to dissociate, $H_2^+ \longrightarrow H + H^+$. Note also that the single electronic states are split into substates corresponding to different energies of mutual vibration of the two nuclei.

These states of H_2^+ are σ states, having axial symmetry. Molecular orbitals can also be antisymmetric across a plane, analogous to the *p* states of the hydrogen atom in which the wave function is positive on one side of a plane through the center and negative on the other side:

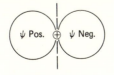

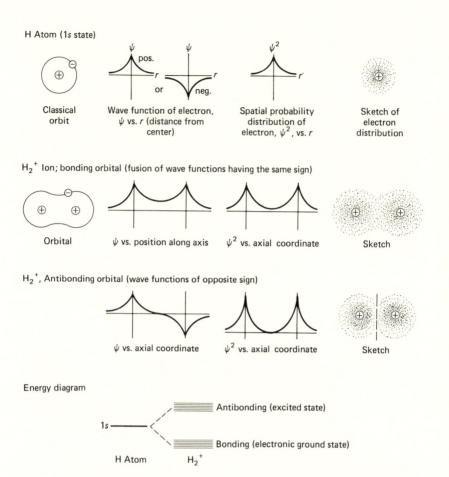

H Atom (1s state)

Classical orbit | Wave function of electron, ψ vs. r (distance from center) | Spatial probability distribution of electron, ψ^2, vs. r | Sketch of electron distribution

H_2^+ Ion; bonding orbital (fusion of wave functions having the same sign)

Orbital | ψ vs. position along axis | ψ^2 vs. axial coordinate | Sketch

H_2^+, Antibonding orbital (wave functions of opposite sign)

ψ vs. axial coordinate | ψ^2 vs. axial coordinate | Sketch

Energy diagram

1s ——— Antibonding (excited state)

Bonding (electronic ground state)

H Atom H_2^+

Fig. 4.2. Sketches of electron distributions in a hydrogen atom and in the molecular ion H_2^+, showing the more stable "bonding" configuration and the less stable "antibonding" case for H_2^+. Individual wave functions ψ are added algebraically when the atomic orbitals fuse to form molecular orbitals; the spatial probability function is given by ψ^2. The energy diagram shows how the changes of stability correspond to changes of energy, and also how the individual electronic states in H_2^+ are split into substates by nuclear vibrations. In this example the spherically symmetric s state of H gives rise to axially symmetric σ states in H_2^+.

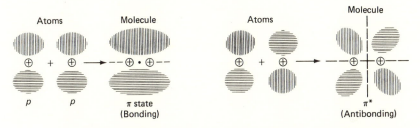

Fig. 4.3. A sketch showing two ways in which atomic p orbitals can fuse when the atoms are together in a molecule. Regions where ψ is positive are shaded vertically, negative regions horizontally. Planes where $\psi = 0$ (zero probability of finding the electron) are shown by dashed lines.

The fusion of such orbitals generates the planar antisymmetry of π orbitals in molecules, as illustrated in Fig. 4.3. Here the excited (antibonding) state is denoted π^*, and a transition from the lower-energy π state to the higher-energy π^* state is called a $\pi\pi^*$ transition.

Consider two atoms, each with an electron in a p orbital, interacting to generate π orbitals. The electrons become the common property of both atoms, and a covalent bond has been formed. Both electrons can occupy the lower (bonding) π orbital if their spins are antiparallel, as allowed by the Pauli Exclusion Principle. A $\pi\pi^*$ transition can then be represented in either of two ways: by an electron energy-level diagram or by a transition diagram, as shown in Fig. 4.4. The energy-level diagram has descriptive utility, as it shows the fates of particular electrons in a transition, but it is inexact in two ways. First, it suggests that the transition involves only the promotion of a single electron to an orbital of higher energy. Second, it is drawn as if the energy levels were not changed by their state of occupancy. These are rough approximations. An electronic transition involves mainly a change in the wave function (and energy) of one electron, but there are lesser changes in all the other components of the molecule. The transition diagram is formally correct; the ground and excited state energies pertain to the entire molecule.

The pattern of π and π^* orbitals, shown in Fig. 4.3 for two atoms, can be extended to a larger number. In benzene there are six carbon atoms in a closed chain of alternating single and double bonds. Each carbon atom contributes one p electron to a pattern that extends over all six atoms. Instead of two distinct orbitals (π and π^*) there are six, resulting from different symmetry patterns of the signs of ψ in adjacent atoms. The six electrons, in pairs with opposed spins, normally occupy the three of these orbitals having the least energy (the greatest bonding character). These are then designated π, and the three higher orbitals, which are normally vacant, are called π^*. With three occupied and three vacant orbitals the pattern of possible transitions is complex. In general, an extended system of π orbitals in an organic molecule

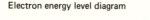

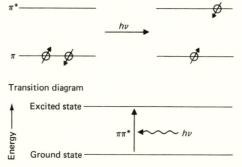

Fig. 4.4. Two ways to represent the excitation of a molecule in a $\pi\pi^*$ transition. The energy-level diagram incorporates two approximations that are discussed in the text. In the transition diagram the energy states pertain to the molecule as a whole.

coincides with a pattern of conjugated (alternating single and double) bonds, with electrons delocalized over the entire pattern. The more extensive the system of conjugated bonds is, the greater the number of distinct symmetry patterns, and the more numerous are the π and π^* orbitals. The energy gaps between successive orbitals are smaller, so that transitions between them involve quanta of less energy or greater wavelength. As an example, various carotenoid pigments have varying numbers of conjugated double bonds in the carbon skeleton. As this number increases, the maxima in the absorption spectrum move to greater wavelengths. For those with 7 or fewer conjugated double bonds the absorption maxima lie in the violet and near ultraviolet; those with 9 to 13 conjugated double bonds absorb in the range 450–550 nm.

In chlorophylls (see Fig. 1.1) a pattern of conjugated double bonds pervades the entire tetrapyrrole ring. Orbitals are closely spaced in energy, and the wavelengths of $\pi\pi^*$ transitions extend well into the near infrared, as can be seen in Fig. 2.11. Further details of the optical spectroscopy of chlorophylls will be considered in Section 7.2.

Molecules also contain electrons that are largely confined to single atoms, in so-called n orbitals. In an "$n\pi^*$" transition an electron moves from a localized n orbital to a delocalized π^* orbital; such a transition involves considerable displacement of charge.

The foregoing descriptions are idealized; the wave function of an electron can usually be described better as a hybrid between n, σ, and π orbitals. But the electron orbitals that spread over extensive regions of conjugated double bonds are predominantly π, and the $\pi\pi^*$ transitions give the most intense bands in optical absorption spectra. These radiative transitions are intense

Table 4.1. *Measures of quantum energy*

Wavelength, λ (nm)	Wave number, k (cm^{-1})	joule per quantum	eV per quantum	kcal per einstein
1240	8×10^3	1.6×10^{-19}	1	23
870	11.5×10^3		1.42	33
680	14.7×10^3		1.82	42

because the redistribution of charge has strong dipole character, and there is good overlap between the ground and excited state wave functions.

4.3   Quantum energy and the energy of oxidation and reduction

The energy of a quantum is $h\nu$ or hc/λ. With $h = 6.6 \times 10^{-34}$ joule sec and $c = 3 \times 10^{10}$ cm sec^{-1}, and with λ expressed in centimeters, the quantum energy in joules is $2 \times 10^{-23}/\lambda$.

Aside from joules, other useful measures of quantum energy are electron volts, calories, and wave number. An electron volt is the energy gained by an electron when it is accelerated through a difference in electric potential of one volt; 1 eV equals 1.6×10^{-19} joule. Wave number, expressed by the symbol k, is defined as the reciprocal of the wavelength in centimeters. It is therefore proportional to quantum energy; E (joule) = $2 \times 10^{-23}k$ for one quantum.

Chemists usually deal not in number of molecules but in gram-moles, and in photochemical processes we are interested in the stoichiometry between quanta and molecules. It is therefore useful to introduce a quantity of light, called an einstein, analogous to a gram-mole:

$$1 \text{ einst} = 6.03 \times 10^{23} \text{ quanta} \tag{4.1}$$

A photochemical reaction that converts one molecule per quantum then converts one gram-mole per einstein. Chemists also prefer to express energy in calories; 1 cal = 4.18 joule.

The foregoing relationships are summarized in the Table 4.1. The wavelengths 680 and 870 nm are approximately those of the long wave absorption maxima of Chl a and Bchl a in vivo, in most organisms that contain these pigments.

A useful formula for the energy of 1 einst of light is:

$$E = \frac{28.5 \times 10^3}{\lambda} \text{ kcal einst}^{-1} \text{ with } \lambda \text{ expressed in nanometers} \tag{4.2}$$

Let us now relate these expressions of quantum energy to the energy of oxidation and reduction.

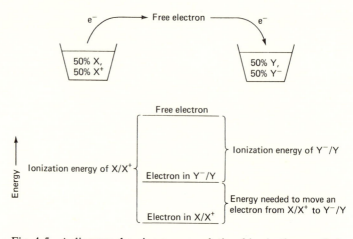

Fig. 4.5. A diagram showing energy relationships in the coupled reactions $X \longrightarrow X^+ + e^-$ and $Y + e^- \longrightarrow Y^-$. (See the text for details.)

To understand the energetics of oxidation and reduction, imagine two vessels containing solutions of hypothetical "redox compounds" such as the X and Y of Reactions 1.7 and 1.8, Chapter 1. The first vessel contains X and X^+ in equal proportions, and the second contains equal parts of Y and Y^-, as shown in Fig. 4.5. Now imagine that a single electron is removed from the first vessel, becoming a free electron. This requires an amount of work equal to the ionization energy (or electron affinity) of the system X/X^+. In this act one X is converted to X^+ as in Reaction 1.7, but the vessel contains so many molecules that the ratio of X to X^+ is not changed appreciably. If the free electron is then allowed to "fall into" the vessel containing Y/Y^-, it releases energy equal to the ionization energy of Y/Y^- and converts one Y to Y^-. The entire transaction requires an input of energy equal to the difference in the ionization energies of X/X^+ and Y/Y^-; this can be expressed in electron volts. This amount of energy has been stored in a pair of processes amounting to $X + Y \longrightarrow X^+ + Y^-$.

Every redox couple such as X/X^+ or Y/Y^- can be placed on a scale of energy in the way suggested by Fig. 4.5, with the energy of a free electron set arbitrarily at zero in the convention used by physicists (the biologists' convention, a scale of oxidation–reduction potentials, will be defined later). Of course, if the proportions of reduced and oxidized forms ($X:X^+$ or $Y:Y^-$) were altered, the ionization energies would change. With a greater proportion of X^+ the system X/X^+ would have more attraction for electrons, and the ionization energy would be greater for an electron in this mixture. For purposes of standardization one can list these energies for equal concentrations[4] of oxidized and reduced forms.

In the convention used by biologists, the arbitrary zero of energy is referred not to a free electron but to an electron in the system $H_2/2H^+$ (the "hydrogen electrode") with 1 atm H_2 and 1 M H^+. Systems of greater electron affinity (stronger oxidants) are given more positive values, and those with less electron affinity negative values. A large negative value implies large energy of the labile electron, closer to that of a free electron. The energy on this scale is called the redox potential[5] and is measured in volts; 1 V means 1 eV for each electron added or withdrawn.

The redox potential is called the midpoint potential, E_m, when the concentrations of oxidized and reduced forms are equal, as in the example of Fig. 4.5. When both forms are at unit activity (1 M for solutes and 1 atm for gases) we speak of the standard midpoint potential, E_m^0. For unequal concentrations the potential is given by a form of the Nernst equation (see any text on chemical thermodynamics):

$$E \text{ (volts)} = E_m^0 - 0.06 \log \frac{[X]}{[X^+]} \tag{4.3}$$

(Throughout this book, logarithms to the base 10 are denoted log; natural logarithms will be denoted ln.) If the stable oxidized and reduced forms of the couple differ by n electrons ($Z_{ox} + ne^- \longrightarrow Z_{red}$) the relationship is:

$$E = E_m^0 - \frac{0.06}{n} \log \frac{[Z_{red}]}{[Z_{ox}]} \tag{4.4}$$

The potential changes by $0.06/n$ volts for every 10-fold change in the ratio of oxidized to reduced forms.

Finally, if an oxidation–reduction reaction involves the binding or release of protons, the energies and equilibria are dictated by the affinities for and concentration of H^+ as well as the electron affinity. For a reaction of the general form

$$Z_{ox} + ne^- + mH^+ \longrightarrow Z_{red}$$

the potential is given by

$$E = E_m^0 - \frac{0.06}{n} \log \frac{[Z_{red}]}{[Z_0][H^+]^m} \tag{4.5}$$

Recognizing that pH is defined as $\log(1/[H^+])$, we can rewrite this as

$$E = E_m^0 - \frac{0.06}{n} \log \frac{[Z_{red}]}{[Z_{ox}]} - 0.06(m/n)(\text{pH}) \tag{4.6}$$

This equation shows how the mechanism of an oxidation–reduction reaction can be explored: By measuring E as it depends on the pH and on the concentrations of Z_{red} and Z_{ox}, one can determine E_m^0, n, and m.

It is often convenient to define a midpoint potential at pH 7, $E_m(7.0)$

instead of the E_m^0 that is based on 1 M H^+ (pH zero). Then Equation 4.6 becomes

$$E = E_m(7.0) - \frac{0.06}{n} \log \frac{[Z_{red}]}{[Z_{ox}]} - 0.06(m/n)(\Delta pH) \qquad (4.7)$$

where ΔpH is the departure from a pH of 7.0.

In the example depicted in Fig. 4.5, imagine that the E_m values are +0.60 V for X/X^+ and +0.36 V for Y^-/Y. Let the concentration of each species (X and X^+ in one vessel; Y^- and Y in the other) be 1 M. If the contents of the two vessels are then mixed, these species will interact so as to bring the redox potential to a common value of E as determined by Equation 4.3:

$$0.60 - 0.06 \log \frac{[X]}{[X^+]} = E = 0.36 - 0.06 \log \frac{[Y^-]}{[Y]}$$

In this approach to equilibrium, every electron gained by X^+ is taken from Y^-, so that the ratio $[X]/[X^+]$ remains equal to the ratio $[Y]/[Y^-]$. Denoting this ratio by R, the equilibrium condition becomes:

$$0.60 - 0.06 \log R = E = 0.36 - 0.06 \log (1/R)$$

Solving this equation, we find $R = 100$ and $E = 0.48$ V. The system comes to equilibrium at a potential of 0.48 V, with (X/X^+) 99% reduced and (Y^-/Y) 99% oxidized.

Now consider a collection of photosynthetic reaction centers in which the photochemical reaction is

$$Chl \cdot A \xrightarrow{h\nu} Chl^+ \cdot A^-$$

with Chl^+ and A^- reverting to Chl and A through "dark" processes as in Reaction 2.3. The two redox couples Chl/Chl^+ and A^-/A are kept apart by the structure of the reaction center and its situation in the membrane, so that each can be at a different potential. If this were not so, energy could not be stored in the photochemical process. Also the separate dark processes, involving secondary electron carriers, allow variations of the stoichiometry: Reaction centers can be in states such as $Chl \cdot A^-$ or $Chl^+ \cdot A$; the ratios of Chl^+/Chl and A^-/A need not be equal. Imagine that the rates of the "light" and "dark" reactions are such that during constant illumination the Chl is kept 99% oxidized, and the acceptor A is kept 91% reduced; $[Chl]/[Chl^+] = 0.01$ and $[A^-]/[A] = 91/9 = 10$. Suppose further that in their natural setting the midpoint potentials of Chl/Chl^+ and A^-/A are 0.4 V and -0.6 V respectively; this is approximately correct for Photosystem 1 in green plants. Then, from Equation 4.3 or 4.4, the couple Chl/Chl^+ is maintained, during the illumination, at a potential of $0.4 - 0.06 \log (0.01)$ or 0.52 V, and the couple A^-/A is maintained at $-0.6 - 0.06 \log (10)$ or -0.66 V. Each photochemical event, driven by a single quantum, moves an electron from +0.52 V to -0.66 V, a gain of 1.18 eV in energy. If the quantum that drives this event has energy of 1.77 eV, corresponding to the "reaction center chlorophyll" in its lowest excited state (long wave absorption maximum 700 nm), a fraction equal to 1.18/1.77 or 67% of the quantum energy has been stored as redox free energy.

4.4   The measurement of light and its uses in photobiology

Consider a uniform beam of light, generated by a point source (a very small lamp filament) at the focus of a lens:

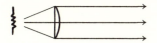

The intensity of the beam can be expressed in terms of the rate at which energy or quanta pass through a surface perpendicular to the beam. This could be measured in $\text{erg cm}^{-2} \text{ sec}^{-1}$, or in W cm^{-2} (1 W = 1 joule sec^{-1} = $10^7 \text{ erg sec}^{-1}$), or in $\text{einst cm}^{-2} \text{ sec}^{-1}$. From Equation 4.2, coupled with the conversion factor $1 \text{ kcal} = 4.18 \times 10^3 \text{ joule}$, we compute $1 \text{ einst} = 12 \times 10^7/\lambda \text{ joule}$, or

$$1 \text{ einst cm}^{-2} \text{ sec}^{-1} = (12 \times 10^7/\lambda) \text{ W cm}^{-2} \qquad (4.8)$$

with λ expressed in nanometers. Conversely,

$$1 \text{ W cm}^{-2} = (8.3 \times 10^{-9}\lambda) \text{ einst cm}^{-2} \text{ sec}^{-1} \qquad (4.9)$$

again with λ expressed in nanometers.

For a beam of light that covers a broad range of wavelengths, one can either apply these equations separately to each narrow interval of wavelength and then combine the results, or else make a rough approximation by choosing a median wavelength in the spectrum of the light. A complete specification would give the energy or quantum flux as a function of wavelength.

Light intensity is also expressed in a variety of units (foot candles, lux, etc.) that are based on the impression of brightness as seen by humans. This is inappropriate in plant science. For example, a beam of light that has been filtered to transmit only the near infrared is invisible to humans and has zero foot candles by definition, but such light is excellent for the growth of photosynthetic bacteria. Translated into practice, tungsten lamps with maximum emission in the near infrared are far better for growing these bacteria than are the much brighter-looking "cool white" fluorescent lamps.

Any relationship between foot candles and energy or quantum flux, for a given light source, involves the spectral composition of the light and the spec-

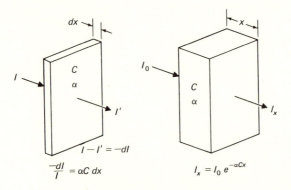

Fig. 4.6. Differential and integrated forms of Beer's Law for the absorption of light by a substance of concentration C and absorption coefficient α. (See further development in the text.)

tral sensitivity curve of a "normal" human observer. If illumination is to be quoted in foot candles, one should at the very least specify the type of lamp, the way in which it is operated, and the type of meter with which foot candles have been measured. A photometer that is calibrated in foot candles is geared to a particular type of illumination, such as "daylight," and will give incorrect readings if used with light having some other spectral distribution.

A "standard candle" gives the same sensation of brightness as a lamp emitting light of wavelength restricted to 555 nm (where the sensitivity of human cone vision is greatest) at a power of $\frac{1}{50}$ W.

Having specified ways to measure light intensity, we shall consider how measurements of the absorption of light are used in photobiological research. The topics to be discussed are absorption spectrometry and the determination of quantum efficiencies and action spectra.

We shall first assign a quantitative measure to the intensity of absorption, as displayed in absorption spectra such as Fig. 2.11. Imagine a thin layer of material such as a solution of molecules, so dilute that the molecules do not shade each other. Let a beam of monochromatic light of intensity I pass perpendicularly through the layer as shown in Fig. 4.6 (left). The fraction of quanta absorbed by the beam is $-dI/I$, the minus sign signifying that the change of intensity dI is itself negative. This fraction is proportional (Beer's Law) to the amount of light-absorbing matter intercepted by the beam, which in turn equals the concentration of light-absorbing substance times the thickness of the layer:

$$-dI/I = \alpha C \, dx \tag{4.10}$$

where C is the concentration of light-absorbing molecules, dx is the thickness, and α is a constant that specifies the intensity or probability of absorption.

This constant is a function of the type of material and the wavelength of the light; the absorption spectrum of Fig. 2.11 could be rescaled as a plot of α versus wavelength.

If Equation 4.10 is integrated over a finite thickness, from zero to x, we obtain the relationship shown on the right in Fig. 4.6:

$$I_x = I_0 e^{-\alpha C x} \tag{4.11}$$

Taking logarithms, converting to base 10 (with $\ln(y) = 2.3 \log(y)$), and rearranging gives the quantity defined as absorbance, a:[6]

$$a = \log(I_0/I_x) = \epsilon C x \tag{4.12}$$

where the "extinction coefficient" ϵ equals $\alpha/2.3$. Since a is dimensionless, the units of ϵ are reciprocal to those of Cx: if C is in molar concentration and x in centimeters, ϵ has the units M^{-1} cm^{-1}.

Equation 4.12 has the form commonly used in spectroscopic chemical analysis. If ϵ has been determined for a certain wavelength, the absorbance at that wavelength allows us to compute the concentration of the substance in question. The virtue of absorbance, as a measure of light absorption, is that it is linear with concentration and additive for a mixture of light absorbers: For a mixture of substances at concentrations $C_1, C_2, \ldots$, with extinction coefficients $\epsilon_1, \epsilon_2, \ldots$,

$$a = (\epsilon_1 C_1 + \epsilon_2 C_2 + \cdots) x \tag{4.13}$$

A pigment with intense absorption typically has $\epsilon = 10^4$ to 10^5 M^{-1} cm^{-1} at a peak in the absorption spectrum. For chlorophylls the peak value (at the long wave absorption maximum) is about 10^5 M^{-1} cm^{-1} or 100 mM^{-1} cm^{-1}.

Returning to Equation 4.10 for a moment: if C is expressed in molecules per cm^3 and dx in centimters, α has units of square centimeters and corresponds to the effective cross section of the molecule for capture of a quantum. A value of $\epsilon = 100$ mM^{-1} cm^{-1} is equivalent to $\alpha = 4 \times 10^{-16}$ cm^2, as one can show by changing units from gram-moles to molecules. This comparison is useful when we wish to compute the rate at which single molecules are absorbing quanta in a light beam. If the intensity of the beam equals I quanta cm^{-2} sec^{-1}, the rate of absorption is αI quanta per molecule per second.[7]

A chemical reaction can be monitored spectroscopically if, at a certain wavelength, the absorbance of the products differs from that of the reactants. An example of interest in photosynthesis is the oxidation of cytochrome. A cytochrome of the c type has a sharp absorption peak near 550 nm in its reduced form but not in its oxidized form, as sketched in Fig. 4.7. Reduction of the oxidized cytochrome raises the value of ϵ at 550 nm from about 5 to 25 mM^{-1} cm^{-1}. We therefore assign a differential extinction coefficient $\Delta\epsilon = 20$ mM^{-1} cm^{-1} to the process of cytochrome reduction, and conversely $\Delta\epsilon = -20$ mM^{-1} cm^{-1} for oxidation of the reduced cytochrome. If the reaction is

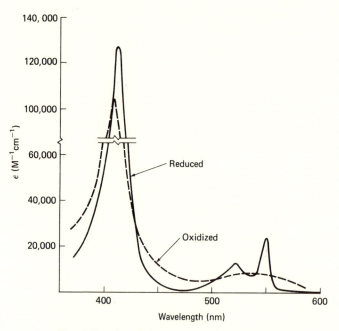

Fig. 4.7. Absorption spectra of mammalian cytochrome c in its oxidized and reduced forms.

observed to give a change of absorbance Δa with path length x, we infer (extending Equation 4.12) that the conversion amounts to a change in concentration:

$$\Delta C = \Delta a / x\, \Delta \epsilon \tag{4.14}$$

We have seen that when the absorption of light is used in chemical analysis, absorbance is a useful measure because it is proportional to the concentration of the light absorber. But when light is used as a photochemical agent we wish to know the rate at which quanta are being absorbed so as to relate this to the rate of the photochemical reaction. If a beam of light of known intensity is incident on a sample, we must determine what fraction of the incident quantum flux is absorbed. Referring again to Fig. 4.6 (right side), the fraction of light transmitted is:

$$T = I_x / I_0 \tag{4.15}$$

and the fraction absorbed, A, equals $1 - T$. The quantities A and T can be computed from the absorbance, and vice versa. Comparing Equations 4.12 and 4.15,

$$a = \log (I_0 / I_x) = -\log T \tag{4.16}$$

An instrument for measuring small changes of absorbance usually gives its information in the form of small changes in the intensity of transmitted light. To convert this information to Δa, we can derive a formula by differentiating Equation 4.16: $da = -d(\log T) = -d(\ln T)/2.3 = -(1/2.3)(dT/T)$. If $\Delta a \ll 1$, this relationship gives a good approximation:

$$\Delta a \approx -\frac{1}{2.3}\frac{\Delta T}{T} \qquad \text{if} \quad \Delta a \ll 1 \tag{4.17}$$

If the absorbance itself is small, we can derive another useful formula by noting that $\ln(1+x) \approx x$ if $x \ll 1$. It follows that $-\log T = -\log(1 - A) = -(1/2.3)\ln(1 - A) \approx -(1/2.3)(-A) = A/2.3$, if $A \ll 1$. Then from Equation 4.16,

$$a \approx A/2.3 \qquad \text{if} \quad a \ll 1 \tag{4.18}$$

Thus with a weakly absorbing sample, a is proportional to A and a spectrum of absorbance vs. wavelength has the same form as a spectrum of "fraction absorbed" vs. wavelength. At the other extreme, with a very concentrated sample the shape of a vs. λ is the same as in a dilute sample of the same material, but the fraction absorbed is close to 100% at all wavelengths and the spectrum of A vs. λ is relatively featureless.

In a photochemical system the quantum efficiency can be measured by using the arrangement shown in Fig. 4.8. The sample, a solution or suspension of the material to be tested, is monitored by a weak beam of wavelength λ_m, chosen so that the reaction being studied will yield a sufficiently large change of absorbance with a known value of $\Delta\epsilon$. While a at λ_m is being recorded, a much stronger beam (the actinic beam) is applied to elicit the photochemistry. The system that detects the measuring light is shielded from stray actinic light by the perpendicular arrangement of the two beams and by the use of complementary color filters, f_m and f_a, such that the actinic light transmitted by f_a is blocked by f_m. A constant intensity of actinic light is applied abruptly, and the resulting rate of photochemistry is determined from the initial slope of a record of a (at λ_m) vs. time (see Fig. 4.8). The incident intensity I_0 of the actinic beam is measured, and also the transmitted intensity I_x. The fraction of incident light absorbed by the sample, A, equals $(I_0 - I_x)/I_0$. In order to correct for light reflected by the sample cell and absorbed by the solvent, the quantity I_0 can be taken as the light transmitted by a sample cell containing solvent but lacking the photochemically active material. The rate at which actinic light is absorbed by the active material is then $I_0 A$ or $I_0 - I_x$. This should be expressed as quanta or einsteins per second, per square centimeter of the actinic beam projected onto the sample.

The computation of quantum efficiency in this system is simplest if we imagine that the sample is a cube 1 cm on each side, with the actinic beam covering one face uniformly and with a path of 1 cm for the measuring beam.

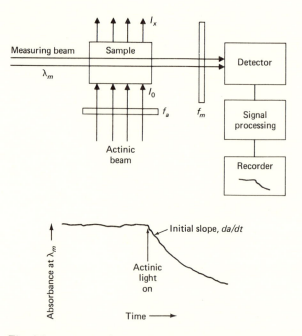

Fig. 4.8. A system for measuring light-induced absorbance changes and the quantum efficiency of a photochemical process. (For details see the text.)

Then the rate of absorption of actinic light is $I_0 A$ (einst cm^{-2} sec^{-1}) in 1 cm^3 of the sample, or $I_0 A$ einst cm^{-3} sec^{-1}. The rate at which material is being converted is $dC/dt = (da/dt)/\Delta\epsilon$ from Equation 4.14 with $x = 1$ cm. If $\Delta\epsilon$ has units M^{-1} cm^{-1}, then dC/dt is in M sec^{-1}, and in the 1-cm cube, material is being converted at the rate $10^{-3} (dC/dt)$ moles cm^{-3} sec^{-1}, or $10^{-3} (da/dt)/\Delta\epsilon$ moles cm^{-3} sec^{-1}. The quantum efficiency, denoted ϕ_p, is then the rate of conversion divided by the rate of light absorption: (moles cm^{-3} sec^{-1})/ (einst cm^{-3} sec^{-1}):

$$\phi_p = \frac{10^{-3} \, da/dt}{I_0 A \, \Delta\epsilon} \tag{4.19}$$

If the sample has dimensions x along the axis of the actinic beam, y along the path of the measuring beam, and height z, and if the actinic beam encompasses the height z and the width y, a similar analysis will show that the quantum efficiency is given by

$$\phi_p = \frac{10^{-3} (x/y) \, da/dt}{I_0 A \, \Delta\epsilon} \tag{4.20}$$

with I_0 in einst cm^{-2} sec^{-1} and $\Delta\epsilon$ in M^{-1} cm^{-1}.

The quantum efficiency measured in this way will be in error if the sample is too dense. Our treatment incorporates the assumption that the part of the sample traversed by the measuring beam is representative of the entire sample exposed to the actinic beam, in terms of the rate of quantum absorption. This is not so, because of the exponential attenuation of light in penetrating the sample (Equation 4.11). In the extreme case of a very dense sample, nearly all the actinic light is absorbed in a thin layer of material at the face on which the actinic beam impinges. The rate of quantum absorption near the center of the cell, where the measuring beam passes through, is then very small. There is little photochemical change in the region sampled by the measuring beam, and the quantum efficiency is underestimated. If the sample absorbs no more than about 50% of the actinic light, and if the axis of the measuring beam passes through the center of the cell, this error is less than about 5%; the central part of the cell gives a reasonably accurate measure of the average rate at which actinic light is absorbed throughout the sample cell. For greatest accuracy one should repeat the measurement with progressively more dilute samples, within the bounds of practicality, and extrapolate the data to infinite dilution.

A powerful technique in photobiology is to analyze the response of the system as a function of the wavelength of the light; that is, to measure an action spectrum. In simple photochemical systems this is usually a trivial exercise: The variation with wavelength should, and generally does, reflect the absorption spectrum of the sensitizing pigment. The same should be true in photobiology, but because the system is usually more complicated, there is more to be learned. There may be several pigments, some effective and others ineffective for the phenomenon under study. The effective pigment(s) might be present in such small amounts as to be invisible. In these cases an action spectrum, or plot of "effectiveness" of the light versus its wavelength, may reveal what pigment is involved or may even show the existence of a hitherto undiscovered pigment.

The first problem in determining an action spectrum is to define and quantify the response. Often this is easy, as with the rate of oxygen evolution by an illuminated leaf, or the angle through which a sunflower has bent in facing toward a light source. In some cases the response is harder to quantify, but a single *standard response* can be described, as with the first perceptible reddening of a patch of skin in an experiment on sunburn.

A simple procedure would be to measure the response in question at a succession of wavelengths, taking care to preserve equal light intensities (quanta cm^{-2} sec^{-1}) at all wavelengths, and then to plot the size of the response versus wavelength. This method cannot be used unless a succession of graded responses can be defined. Even if the response is readily made quantitative, as with a photochemical rate, its magnitude might not be simply proportional to light intensity. A more complicated relation between response and light intensity will then be translated into a distortion of the shape of the action spectrum: Peaks will be "unnaturally" flattened or exaggerated relative to the absorption spectrum.

These difficulties are avoided by following a slightly different procedure. A single standard magnitude is chosen for the response, and at each wavelength the light intensity is found that will produce that response.[8] Less light will of course be needed at the wavelengths of greater effectiveness. The reciprocal of the intensity needed for a standard response is thus a measure of the effectiveness of the light; this is plotted against wavelength to produce an action spectrum. The rationality of this procedure will now be defended for a case where the response is well defined – for the photochemical oxidation of cytochrome in an alga.

Equation 4.19 can be rewritten in the form

$$\frac{1}{I_0} = \frac{1}{R} A \phi_p \qquad (4.21)$$

where the quantity R (response) is substituted for $10^{-3}(da/dt)/\Delta\epsilon$ or moles cm^{-3} sec^{-1} of photoproduct. We have thus defined the response as the rate at which a product is made. Now suppose that we choose a particular value of R as a standard response and determine at each wavelength the value of I_0 that is needed to give this response. Having stipulated that R is constant, let us suppose for the moment that ϕ_p is also constant (independent of wavelength). Then by Equation 4.21, a plot of $1/I_0$ vs. wavelength (the action spectrum) should have the same shape as a plot of A (fraction absorbed) vs. wavelength (the absorption spectrum). A comparison of these plots is thus a test of whether or not the quantum efficiency ϕ_p actually is constant, or whether it varies with wavelength.

SUGGESTED READINGS

Feher, G. (1970). *Electron Paramagnetic Resonance with Applications to Selected Problems in Biology*. Gordon and Breach, New York.

Rice, F. O., and Teller, E. (1949). *The Structure of Matter*. Wiley, New York.

Clayton, R. K. (1965). *Molecular Physics in Photosynthesis*, Chapters 6 and 7, pp. 66–96. Blaisdell, New York.

Clayton, R. K. (1971). *Light and Living Matter*, Vol. 1, Chapter 3, pp. 64–108. McGraw-Hill, New York. Reprinted in 1977 by R. E. Krieger, Huntington, New York.

Part II

PIGMENT-PROTEIN COMPLEXES IN
PHOTOSYNTHETIC MEMBRANES:
THEIR COMPOSITIONS, STRUCTURES,
AND FUNCTIONS

The photosynthetic apparatus (pigments, photochemical reaction centers, and associated enzymes and coenzymes) is carried in and on protein–lipid membranes that are developed so as to have a large surface area. In plant cells and algae (excluding the blue-green algae that are now classified as Cyanobacteria) these membranes are within the subcellular bodies called chloroplasts. In bacteria the photosynthetic membranes are derived by invagination of the cytoplasmic membrane, and are elaborated into a variety of shapes.

Figure 5.1 shows an electron micrograph of a section through a chloroplast from spinach. Here the photosynthetic membranes have the form of flattened bags called thylakoids. They lie roughly parallel to each other in a lamellar pattern that extends throughout the chloroplast, and in some places they are stacked more closely to form the dense regions called grana. These grana stacks are not isolated bodies; in fact, serial sections suggest that the entire system of thylakoids in one chloroplast forms a syncytium, so that the interiors of all the thylakoids are topologically connected. The lamellae that lie outside the stacked regions are called stroma lamellae. We shall see that the membranes in the grana and stroma regions differ in their fine structure and to some degree in their composition and function. It is important to note that these membranes form a barrier between an enclosed inner region and an outer region. Thylakoid membranes of chloroplasts can be frozen and fractured to expose inner interfaces. These are studded with protein subunits, as shown in the electron micrograph of Fig. 5.2. Details of the structure, composition, and function of thylakoid membranes are described in Chapter 6.

In many types of photosynthetic bacteria, invaginations of the cytoplasmic membrane appear as round vesicles inside the cell. The membrane bounding these vesicles has become differentiated (specialized for photosynthesis) from the parent cytoplasmic membrane, and is called intracytoplasmic membrane. The round vesicular bodies

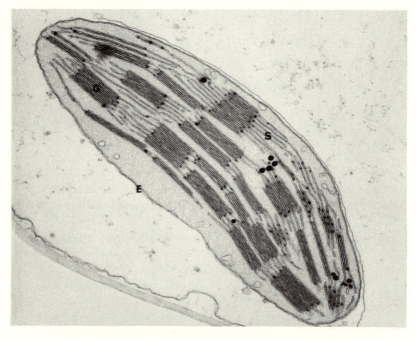

Fig. 5.1. Electron micrograph of a thin-sectioned spinach chloroplast, showing the thylakoids in cross section. The chloroplast is about 10 μm long. Thylakoids are appressed in stacks ("grana," G) which are interconnected by stroma lamellae (S). A double envelope membrane (E) encloses the chloroplast. Courtesy of Dr. C. J. Arntzen, University of Illinois.

can be seen in profusion in Fig. 5.3, a section through a cell of *Chromatium vinosum*. Their derivation as invaginations of the cytoplasmic membrane is especially clear in Fig. 5.4, a section through *Rhodopseudomonas sphaeroides* in which there are relatively few such vesicles. For the most part these round objects are cross sections of invaginations that are still connected to the cytoplasmic membrane. In a single picture such as Fig. 5.4 it is not clear whether any of the vesicles have become detached inside, but a special technique gives information on this point. The bacteria can be treated chemically so as to remove the cell wall that surrounds the cytoplasmic membrane; the resulting naked cells are called spheroplasts. The membrane and its invaginations remain intact, except that the invaginations balloon out. In these spheroplasts there is no indication of smaller interior vesicles.

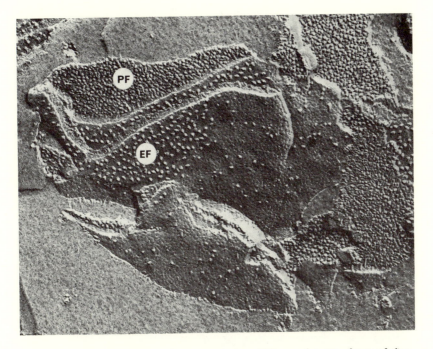

Fig. 5.2. Electron microscope image of freeze-fracture faces of thylakoids isolated from spinach chloroplasts. The planes of fracture follow the hydrophobic interiors of the bilayered membranes of the thylakoids, exposing complementary views (PF and EF) of this interface. The particles on the EF face are about 12–16 nm in diameter; those on PF have diameters about 8–10 nm. For interpretation see Section 6.2; also C. J. Arntzen (1978), *Curr. Top. Bioenerg. 8*, 111–60. Courtesy of Dr. P. Armond, University of Illinois.

If normal photosynthetic bacteria are broken, as by sonic disruption, both the cytoplasmic and intracytoplasmic membranes are fragmented, and the fragments re-form into sealed round vesicles. These can be isolated by differential centrifugation; a suspension of such vesicular membrane fragments from bacteria is commonly called a chromatophore preparation.

Examination of Fig. 5.4 shows that the region outside the cytoplasmic membrane (the periplasm, between the cytoplasmic membrane and the cell wall) is carried into the inside of an invagination. The outside of the cytoplasmic membrane is thus topologically equivalent to the inside of the intracytoplasmic vesicle. The chromatophores derived from broken cells have the same topology as the intracytoplasmic vesicles. This can be tested by the fact that light

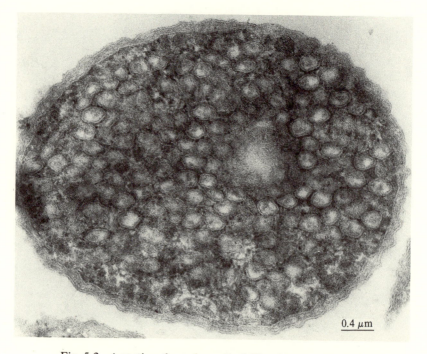

0.4 μm

Fig. 5.3. A section through a cell of *Chromatium vinosum* (larger than average), packed with vesicular structures of intracytoplasmic membrane arising from invaginations of the cytoplasmic membrane. Courtesy of Dr. S. F. Conti, University of Kentucky.

promotes the unidirectional pumping of H^+ ions across the membrane, as a consequence of the photochemistry of photosynthesis. In cells that have lost the exterior cell envelope and cell wall, leaving the entire cytoplasmic membrane intact ("spheroplasts"), H^+ ions are pumped outward, across this membrane from the cytoplasm to the exterior. The topology then suggests that H^+ should be pumped *into* the intracytoplasmic vesicles, from the cytoplasm to the internally trapped periplasm. In suspensions of vesicles isolated from broken cells, light causes the uptake of H^+ by these chromatophores from the external medium. The thylakoids of chloroplasts show similar directionality; light drives H^+ to the interior of the thylakoid.

In some species of photosynthetic bacteria the intracytoplasmic membranes have the shape of tubes; in others the invaginations of the membrane are folded back and forth in sheets to give stacked lamellae. Figure 5.5 shows a section through *Thiocapsa violacea*, with indications of both lamellar and tubular development, although

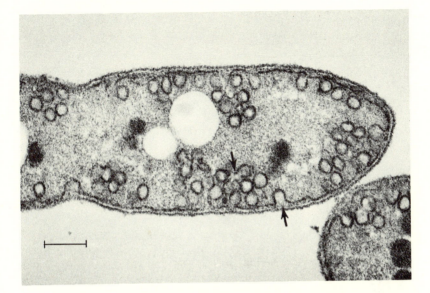

Fig. 5.4. A section through *Rhodopseudomonas sphaeroides* containing relatively little intracytoplasmic membrane. The origin of intracytoplasmic membrane by invagination of the cytoplasmic membrane can be seen at the places indicated by arrows. The marker bar is 0.2 μm long. Courtesy of Dr. G. A. Peters, C. F. Kettering Research Laboratory.

most of the intracytoplasmic membrane looks like round bodies in cross section.

The surfaces of at least some intracytoplasmic membranes show a hexagonal pattern of latticelike fine structure. This has been revealed by a powerful combination of electron microscopy and diffraction analysis that has proved to be extremely useful for showing repetitive or quasicrystalline fine structure in macromolecular assemblies. A photograph of an electron microscope image is used as the source of an optical diffraction pattern. The diffraction pattern is processed to delete "noise" and to enhance contrast, and is then converted back to a primary image (image reconstruction). The final image is like the original one, but with any regularly repeating features enhanced.

In the green photosynthetic bacteria (Fig. 5.6) there are many large vesicles lining the periphery of the cell. These vesicles contain the accessory light-gathering pigment bacteriochlorophyll *c*; their membranes are entirely different from the cytoplasmic membrane.

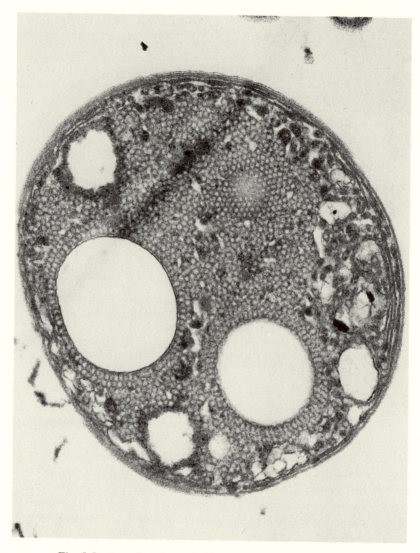

Fig. 5.5. A section through *Thiocapsa violacea*, a purple sulfur bacterium related to *Chromatium*. The diameter of this section is about 3 μm. The densely packed intracytoplasmic membrane shows indications of lamellar and tubular structure near the periphery. Courtesy of Dr. C. C. Remsen, University of Wisconsin.

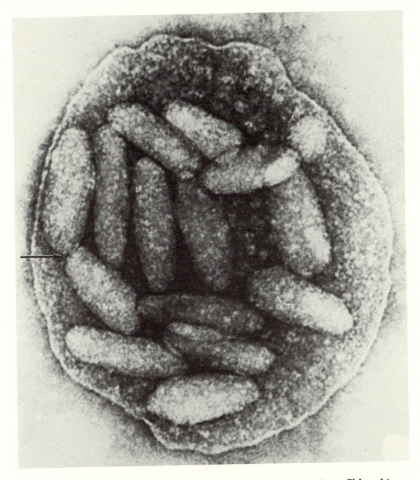

Fig. 5.6. A section through the green sulfur bacterium *Chlorobium limicola*, showing the large vesicles that contain bacteriochlorophyll *c*. Cell diameter about 0.7 μm. Courtesy of Drs. S. C. Holt and R. C. Fuller, University of Massachusetts, and Dr. S. F. Conti, University of Kentucky.

In these bacteria the remainder of the photosynthetic apparatus is probably in the cytoplasmic membrane adjacent to the vesicles, or sandwiched between the cytoplasmic membrane and the vesicles.

In blue-green algae and red algae the chlorophyll-containing photosynthetic membranes are studded with little knobs as shown in Fig. 5.7. These knobs are the phycobilisomes that carry the accessory phycobilin pigments.

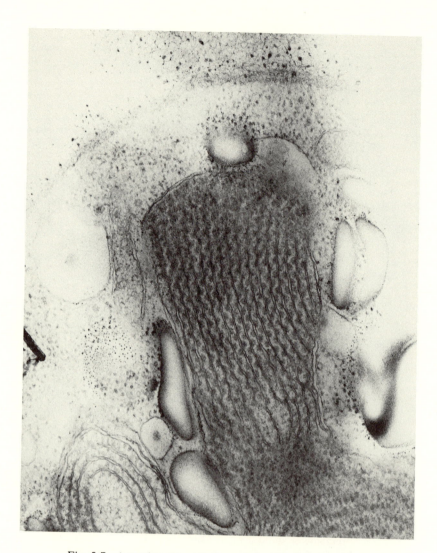

Fig. 5.7. A section through the red alga *Porphyridium cruentum*, showing part of a chloroplast with its lamellar thylakoids. The phycobilisomes can be seen as fuzzy particles on the surfaces of the thylakoids. The section of chloroplast shown here is about 1 X 2 μm; the entire cell is about 5 μm in diameter. Courtesy of Dr. E. Gantt, Smithsonian Institution, and Dr. S. F. Conti, University of Kentucky.

In Part II we shall examine the structures of photosynthetic tissues in relation to energy transfer, photochemistry, and subsequent electron and proton transfer. We shall start with the simpler bacterial system and progress to the more complex arrangements in green plants. Considerable details about the reaction centers of bacteria will be given in Chapter 5; a treatment of the more obscure reaction centers of green plants will be deferred to Chapters 7 and 8.

5 Components of the photosynthetic membranes of bacteria: composition and function in energy transfer and photochemistry

5.1 The photochemical reaction centers

In photosynthetic membranes of bacteria the photochemical reaction centers are distinct pigment–protein complexes. The first direct indication of their existence came when Duysens described the reversible light-induced bleaching of a component absorbing light near 870 nm, in chromatophores (intracytoplasmic membrane fragments) of the photosynthetic bacterium *Rhodospirillum rubrum*. This was manifested as a small (about 2%) light-induced decrease of absorbance in the major long wave absorption band of Bchl *a*. In the dark the band grew back to its former height. A similar change could be induced by chemical oxidation, and the actions of light and chemical oxidants were complementary. If a submaximal change was induced chemically, light could do no more than to induce the remainder of a normal maximal (~2%) change. It appeared that light causes the reversible oxidation of a pigment resembling Bchl *a* in these bacteria, detectable by loss of the absorption band (near 870 nm) of the neutral form of the pigment. This reactive pigment, now certified to be Bchl *a* (see later in this section), is usually called P870 (P for pigment, 870 for the wavelength of maximum absorbance change). In cells and subcellular fractions of various bacteria the bleaching may be centered at wavelengths ranging from 860 to 890 nm; the generic term P870 (or even the abbreviation P) can be retained for convenience. Green plants have the counterparts P700 in Photosystem 1 and P680 in Photosystem 2; these are specialized forms of Chl *a*.

Duysens had no way to decide whether the absorbance change was due to the gross alteration (oxidation) of a specific minor fraction of the Bchl, or to a small decrease in absorption by all the Bchl. But methods were discovered, in my laboratory and in others, to eliminate irreversibly most of the 870 nm absorption band without attenuating the component responsible for the reversible light-induced change. In this way a specific, minor, photochemically active component of the Bchl could be isolated, at least as an entity in the absorption spectrum. One successful method, illustrated in Fig. 5.8, was to expose chromatophores to strong light in the presence of a detergent such as Triton X-100. Another method was to mix the chromatophores with the

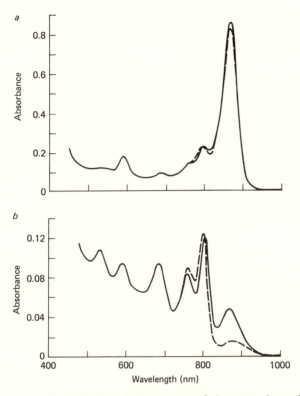

Fig. 5.8. Absorption spectra of chromatophores from carotenoidless mutant *Rhodopseudomonas sphaeroides:* (*a*) before treatment and (*b*) after exposure to strong light in the presence of 1% Triton X-100. The difference between the dashed curve and the solid curve shows the reversible light-induced absorbance change. In (*b*) the absorption band at 680 nm reflects a degradation product, an oxidized derivative of Bchl *a*. The dashed curve shows at 870 nm the presence of some antenna Bchl *a* that had not been degraded by the treatment.

powerful oxidant potassium chloroiridate, K_2IrCl_6. Either of these treatments caused the selective degradation of the antenna Bchl, deleting its long wave absorption band, while leaving most of the photochemically active Bchl intact. This active Bchl was presumed to be a part of the photosynthetic reaction center. Figure 5.8 shows that in addition to the reversibly bleachable 870 nm band, other absorption bands near 760 and 800 nm survived the treatment. We shall see later that these bands, which appear as minor shoulders in the absorption spectrum of the untreated chromatophores (Fig. 5.8, left), also belong to pigments in the reaction center, separate from the antenna Bchl.

The isolation of reaction centers as chemical entities was developed in my laboratory in 1967, and the techniques have since been improved and extended to many species of photosynthetic bacteria. Reaction centers are isolated by breaking the bacteria, exposing the intracytoplasmic membrane fragments to a detergent so as to break down the membranes into components, and then isolating the specific reaction center component by standard biochemical methods for fractionating proteins, such as column chromatography and differential precipitation. The detergent displaces phospholipids from the membrane, allowing some of its components to fall away. These membrane-derived proteins are strongly hydrophobic; their solubility in water requires the continued presence of a detergent.

The best procedure for isolating reaction centers depends on their detailed physical properties, which vary from one organism to another. Reaction centers from *Rhodopseudomonas sphaeroides* are exceptionally stable and are easily isolated[1] from mutant strains that lack colored carotenoids. Accordingly these reaction centers have become favorites objects of study, and our knowledge of their composition and properties has been brought to a high degree of refinement. At this writing reaction centers have been isolated from at least six different species of photosynthetic bacteria, embracing widely different genera. They all have the same general composition and properties, b the details vary. In every case the primary photochemical electron donor is Bchl and the first stable electron acceptor is a quinone, usually ubiquinone (see Fig. 1.4).

Before considering the composition and functioning of these reaction centers in more detail, let us review the evidence that they really are the sites of the primary photochemistry of bacterial photosynthesis. By the mid-1950s it was known from light-induced absorbance changes that light causes the oxidation of two distinct entities in photosynthetic bacteria, cytochromes and Bchl. It became evident that an oxidized cytochrome of the c type is a major photoproduct in living cells or chromatophores exposed to weak light. The bleaching near 870 nm that signals the oxidation of Bchl is seen only after all of the cytochrome has been driven to its oxidized form, for example, in continuous strong light. This suggested to some that the oxidation of cytochrome is part of the primary photochemistry of photosynthesis. In this view the normal photochemistry of photosynthesis involves the transfer of an electron from cytochrome to Bchl, yielding oxidized cytochrome and reduced Bchl (Bchl$^-\cdot$). The oxidation of Bchl was regarded as an aberrant reaction, or perhaps a safety valve to discharge excess light energy harmlessly. Only when the cytochrome is oxidized, and therefore cannot act as primary photochemical electron donor, does Bchl become oxidized. This idea was appealing because Bchl and the prosthetic group (heme) of cytochrome are both tetrapyrroles. Chemists could visualize the transfer of an electron from a π orbital in cytochrome to one in Bchl. But there was no evidence for the formation of Bchl$^-\cdot$, and one could argue with equal force that oxidized P870 ("P$^+$")

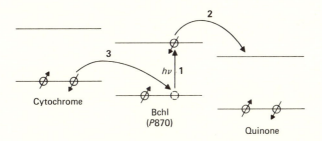

Fig. 5.9. A pictograph showing the movements of electrons in the photochemistry of bacterial photosynthesis. Step 1 is the excitation of P870 (the photochemically active Bchl), step 2 is the reduction of a quinone, and step 3 is the oxidation of a cytochrome. The overall reaction is $Cyt_{red} + Q \longrightarrow Cyt_{ox} + Q^-\cdot$, with oxidized P870 appearing as a transient intermediate because step 2 precedes step 3. The diagram is incomplete, in that bacteriopheophytin acts as an intermediate electron carrier between P870 and the quinone.

is a primary photoproduct, with cytochrome oxidation a closely coupled "dark" reaction: $Cyt_{red} + P^+ \longrightarrow Cyt_{ox} + P$. If this secondary reaction were rapid enough, one would not expect to see any accumulation of P^+ until all of the cytochrome had been driven to its oxidized form.

Figure 5.9 shows how both views can be represented in a single energy level diagram, a crude portrayal of probable events. The first step is the excitation of Bchl (P870); a quantum of light energy promotes an electron to a higher level. This event facilitates subsequent electron transfer events: The electron promoted in P870 can enter a vacant level in the acceptor Q (quinone), and the cytochrome can deliver an electron to the vacancy (created by excitation) in the lower level of P870. The diagram has been arranged so that both of these electron transfer events are exergonic. If the step labeled (2) happens before the one labeled (3), the earliest photochemical products are P^+ and $Q^-\cdot$. The P^+ is then re-reduced to P by receiving an electron from Cyt_{red}. But if step (3) occurs before step (2), the first products of photochemistry are Cyt_{ox} and $Bchl^-\cdot$, with $Bchl^-\cdot$ then delivering its extra electron to Q. Thus the two views, seemingly very distinct, come down to the simple question of which electron transfer, (2) or (3), is the faster.

With the advent of absorbance measurements using pulsed laser excitation, W. W. Parson was able to resolve the issue. In chromatophores of *Chromatium vinosum*, a laser flash of duration less than a microsecond caused the immediate oxidation of P870 as seen by loss of absorbance at 870 nm. The P^+ returned to its reduced form with half-time about 2 μsec, and the absorbance change signifying cytochrome oxidation "grew in" with the same half-time, paralleling the return of P^+ to P. Evidence for the reduction of Q was lacking in these experiments, but it became clear that step (2) precedes step (3). The primary photoproducts are P^+ and (in retrospect) $Q^-\cdot$; the oxidation of cyto-chrome follows when cytochrome is available in its reduced form.

Parson went on to show that in chromatophores of *Chr. vinosum* the quantum efficiency of flash-induced Bchl oxidation is about 90% or greater. Most of the quanta absorbed by the tissue are used in making P^+ (and $Q^-\cdot$).

A final piece of evidence certified that the reaction centers, in which P870 is photo-oxidized to P^+, are indeed the sites of the photochemistry of photosynthesis. This evidence was found in a mutant of *Rp. sphaeroides* isolated by W. R. Sistrom. The mutant (strain PM-8) cannot grow by photosynthesis; it must be grown aerobically, with respiration providing the energy. Cells and chromatophores of this mutant show no sign of the photochemistry of photosynthesis: no light-induced absorbance changes or ESR signals and no variations of the yield of fluorescence during illumination. Attempts to isolate reaction centers from the mutant are negative, and tests with specific antibodies, prepared by injecting rabbits with reaction centers purified from the photosynthetically competent parent strain, show that the mutant lacks the characteristic protein components of the reaction-center particle. Thus in a single mutation the organism has lost the ability to make reaction centers and the ability to perform photosynthesis.

The composition of reaction centers from *Rp. sphaeroides* is illustrated in Fig. 5.10. In each reaction-center particle, four molecules of Bchl and two of bacteriopheophytin[2] (Bpheo) are bound noncovalently (by hydrogen bonds, hydrophobic interactions, and other weak intermolecular forces) to a protein composed of three polypeptides. These polypeptides have been characterized as to molecular weight and amino acid composition. Their apparent weights as determined by electrophoresis through polyacrylamide are 19, 22, and 28 kdalton (a weight of 19 kdalton corresponds to a molecular weight of 19,000), but these values are probably underestimated by about 30%.[3] The three polypeptides are termed L, M, and H for light, medium, and heavy, respectively. The reaction-center protein, containing a high proportion of nonpolar amino acids, is insoluble in water unless a detergent is present. By coating the protein and presenting a hydrophilic outer surface, the detergent makes the system soluble in water.

Each reaction-center particle also contains one iron atom and one or two molecules of ubiquinone (UQ). Only one UQ is essential for photochemistry – the Fe possibly facilitates the transfer of electrons from the first ("primary") UQ to the second; see Section 8.1.

The content of Bchl, Bpheo, and UQ in reaction centers has been determined by extracting a known amount of the reaction centers with a suitable organic solvent and assaying the extracted compounds by their optical absorption.

The reaction-center particle can be defined as a photochemical unit because it is capable of transferring no more than one electron at a time, despite the

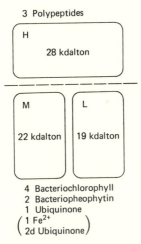

3 Polypeptides

H

28 kdalton

M

22 kdalton

L

19 kdalton

4 Bacteriochlorophyll
2 Bacteriopheophytin
1 Ubiquinone
$\left(\begin{array}{l} 1\ Fe^{2+} \\ 2d\ Ubiquinone \end{array}\right)$

Fig. 5.10. The reaction-center particle isolated from *Rhodopseudo-monas sphaeroides* contains three polypeptides "H, M, and L," weighing 28, 22, and 19 kdalton (i.e., molecular weights 28,000, 22,000, and 19,000), respectively. (Note: These molecular weights, determined by electrophoresis in polyacrylamide, are probably underestimated by about 30%; see the text.) Four molecules of Bchl and two of bacteriopheophytin (Bpheo; like Bchl with 2H in place of Mg at the center) are bound noncovalently to the polypeptides, along with one or two molecules of ubiquinone and one atom of ferrous iron. The H polypeptide can be detached, leaving an "LM particle," which retains the pigments and can perform apparently normal photochemistry (photochemical transfer of an electron from Bchl to the first ubiquinone). The Fe^{2+} and the second ubiquinone are not essential for this photochemistry; Fe^{2+} probably facilitates the transfer of electrons from primary to secondary ubiquinone.

presence of four Bchl molecules. A single flash of light can drive one electron from Bchl to UQ, and the reaction center cannot repeat this act until the photoproducts, oxidized Bchl and reduced UQ, have been neutralized by further electron transfer (Reaction 2.3) or back-reaction. We shall see in Sections 7.2 and 8.1 that two of the four Bchls share the function of photochemical electron donor, and that one of the two Bpheos acts as an intermediary electron carrier between Bchl and UQ. Following conventional terminology we shall retain the terms P and P^+ for the neutral and oxidized states of the "special pair" of Bchl molecules that cooperate in donating an electron (the two states of the special pair are sometimes denoted B_2 and $B_2^{+}\cdot$). The intermediary carrier, one Bpheo (with perhaps some involvement of the "nondonor" Bchls), is denoted I in its neutral state and I^- in its reduced state.

$$PIQ \xrightarrow{h\nu} P^*IQ \longrightarrow P^+I^-Q \longrightarrow P^+IQ^- \qquad (5.1)$$

The functions of the other two Bchls and the other Bpheo in each reaction center are not well understood; this ignorance is reflected in the temporary designation "voyeurs" for these seemingly superfluous pigments.

The H subunit of the reaction center can be split off under mildly denaturing conditions, leaving the combination of L and M intact. The resulting "LM particle" retains the four Bchls, the two Bpheos, and the primary UQ, and can perform apparently normal photochemistry. The LM particle is therefore the smallest known entity in the photosynthetic membrane that retains the capacity for "normal" photochemistry.

Structural relationships between the H, M, and L subunits, including their deployment in the photosynthetic membrane and the sites of interaction with electron donors and acceptors (cytochromes and quinones), are discussed in Section 8.1, with a model shown in Fig. 8.3.

An absorption spectrum of reaction centers from *Rp. sphaeroides*, and two spectra of light-induced absorbance changes (measured under different conditions), are shown in Fig. 5.11. In the absorption spectrum (*a*) the bands near 870, 800, and 600 nm are due to Bchl; those near 760 and 535 nm are due to Bpheo. The band near 365 nm is a composite of unresolved bands of Bchl and Bpheo, and the band at 280 nm is due mainly to the aromatic amino acids in the protein. At the 800 nm peak the extinction coefficient (Section 4.4) ϵ equals 300 mM^{-1} cm^{-1}.

A fluorescence spectrum of these reaction centers shows a band near 900 nm, corresponding to the absorption band at 870 nm. Fluorescence bands between 700 and 800 nm are also present to greater or lesser degrees, but these are attributable to contamination of the reaction centers by extraneous pigments, principally detached Bchl and Bpheo.

Figure 5.11*b* shows the spectrum of reversible light-induced changes in reaction centers to which nothing has been added. The greatest feature is the negative band centered near 870 nm, representing the complete bleaching of the 870 nm band. There is a similar bleaching centered near 600 nm, where Bchl has an absorption band. The positive bands near 435 nm and 1250 nm, and the broad absorbance from about 460 to 570 nm, are properties of the oxidized special pair of Bchl, P$^+$. Taken together, all these features show the conversion of P to P$^+$ by the transfer of a single electron to UQ. The bleaching centered at 275 nm is due mainly to the reduction of UQ, Q $\rightarrow$ Q$^-$. Other features attending this reaction are smaller and are submerged in the larger changes due to P $\rightarrow$ P$^+$. When illumination is stopped, these light-induced changes decay owing to a back-reaction between P$^+$ and Q$^-$. The half-time of this restoration ranges from about 60 msec to several seconds, depending on whether the electron is situated on the primary UQ or has moved on to a secondary one.

The features in Fig. 5.11*b* centered around 800 nm suggest that the absorption band of Bchl at 800 nm has shifted to a lesser wavelength, but this is

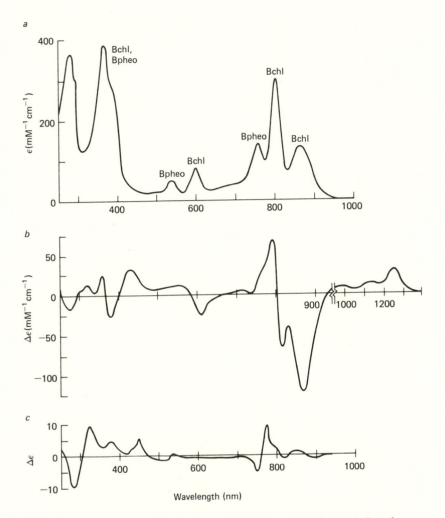

Fig. 5.11. Absorption spectrum (*a*) and spectra of light-induced absorbance changes (*b*, *c*) of reaction centers isolated from carotenoidless mutant *Rp. sphaeroides*. Trace *b*, reaction centers alone. Trace *c*, reaction centers plus an electron donor to prevent the accumulation of oxidized bacteriochlorophyll during illumination. Changes due to oxidation of the donor have been discounted. These spectral features and the corresponding photochemical processes are discussed in the text. In *a*, bands due to bacteriochlorophyll and to bacteriopheophytin are labeled Bchl and Bpheo, respectively. The ordinate is the millimolar extinction coefficient ϵ or differential extinction coefficient $\Delta\epsilon$ (see Section 4.4), based on the molarity of reaction centers regarded as macromolecules. Note the different scales in *a*, *b*, and *c*.

only part of the explanation, as we shall see in Section 7.2. A similar "band shift effect" can be discerned also around 570–600 nm, superimposed on other changes in this region. The ripple around 760 nm could be caused by a shift to greater wavelengths of the 760 nm absorption band of Bpheo. The changes near 370 nm could be due to shifts of bands of either Bchl or Bpheo. These effects are attributable partly to the influence of local electric fields (generated by the photochemical charge separation) on the absorption spectra of the pigments. They will be examined in greater detail in Sections 7.2 and 8.1.

In the reaction centers described by Fig. 5.11b the photochemical products are P^+ and Q^-. If these reaction centers are mixed with an excess of an electron donor such as reduced cytochrome, P^+ does not accumulate as a photoproduct because it reacts rapidly with the donor:

$$Cyt_{red} + P^+IQ^- \longrightarrow Cyt_{ox} + PIQ^- \tag{5.2}$$

Light-induced absorbance changes will show the oxidation of cytochrome and the reduction of UQ, and if the spectra can be corrected to discount the oxidation of cytochrome, the remaining spectrum will pertain to the conversion of PIQ to PIQ$^-$. This is shown in Fig. 5.11c, from experiments using donors other than cytochrome so as to avoid the large corrections associated with cytochrome oxidation. The negative band at 275 nm and the positive bands in the region 320 to 450 nm reflect the conversion of UQ to UQ$^-$·. Other features reflect band shifts of Bpheo and Bchl.

Finer details in these spectra, relating spectral features to specific molecules ("active" and "voyeur") of Bchl and Bpheo, will be discussed in Sections 7.2 and 8.1.

In general, the reaction centers derived from carotenoidless mutant *Rp. sphaeroides* are representative of reaction centers from a wide variety of bacterial sources. Only the details vary. For example, if reaction centers are isolated from organisms that contain colored carotenoids (such as the wild type strain of *Rp. sphaeroides*, from which the carotenoidless mutant was derived), one molecule of carotenoid is bound to each reaction-center particle. The primary quinone in reaction centers from some species is menaquinone, although the secondary quinone is always UQ. In species that contain Bchl b as antenna pigment, the Bchl and Bpheo in the reaction centers are also of the b type. As a result the absorption bands of Bchl are all found at greater wavelengths; for example,[4] at 980 and 830 nm instead of at 870 and 800 nm. Finally, the pattern of protein components varies. In most cases there are three distinct polypeptides in the approximate range 20–30 kdalton, as with *Rp. sphaeroides*, but in some cases only two slightly heavier ones are found.

Reaction centers of the green photosynthetic bacteria are in a separate category and have not yet been purified. The electron donor is Bchl a absorbing near 840 nm (P840). The first stable electron acceptor is not a quinone; it is a much stronger reductant, with midpoint potential perhaps as negative as -0.6 V. For the primary quinone in the majority of photosynthetic bacteria the midpoint potential $(Q/Q^{-}\cdot)$ is about -0.2 V.

5.2 Digression: relationships between absorption, fluorescence, and photochemistry

We have seen (Section 2.1) that molecules can undergo radiative transitions between ground and excited states, and that the probability of such an event is governed by the pattern of redistribution of charge during the transition. For high probability this pattern must correspond to a transitory dipole oscillation, and the electron distribution must not change too radically: The ground and excited state wave functions must overlap well. In classical terms one can say that the electromagnetic wave forces the dipole oscillation and thus initiates the transition; for strong absorption or emission (high transition probability) the frequency of the light must match the difference in energy between ground and excited states ($\Delta E = h\nu$). These considerations apply equally to upward and downward transitions, but there are two types of downward (emissive) transition, spontaneous and induced.

An induced downward transition is the counterpart of an upward transition, being driven by external radiation, but with a quantum emitted rather than absorbed:

The probabilities of these upward and induced downward transitions are equal and proportional to the intensity of electromagnetic radiation impinging on the molecule. There are also spontaneous downward transitions, for which the probability is unrelated to the electromagnetic radiation around the molecule.

Under constant illumination a collection of molecules will reach a steady state in which the rates of upward and downward transitions are equal. At "ordinary" light intensities, as produced by lamps other than lasers, the probability of a spontaneous downward transition is far greater than that of an induced (upward or downward) transition. Then in a steady state most of the molecules are in their ground states, and spontaneous emission (fluorescence) predominates over induced emission. At high-enough light intensity the induced emission becomes predominant, and spontaneous emission is negligible in comparison. As the intensity approaches infinity, a collection of molecules will approach a steady state in which half are excited and half are in the ground state, with equal rates of induced upward and downward transitions. Thus the excited state population can never exceed 50% in a steady state where only direct transitions between a ground state and a single excited state are involved.

If the excited state can be reached indirectly, as from a higher excited state, the collection of molecules can attain a "population inversion" with more than half in the excited state:

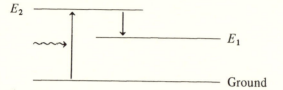

The radiation that drives molecules into the higher state E_2, and thence to the lower E_1, is not of the right frequency to stimulate a transition from E_1 to ground, and the majority of the population can thus be "pumped" into E_1. This can be an explosive situation. A single spontaneous transition from E_1 to ground will produce a quantum of the right frequency to stimulate an induced transition in another molecule. The induced transition yields two quanta for one, and given the right geometry, each of them can yield two quanta, and so forth. In a chain reaction, a cascade of induced transitions, all the energy stored in the population of excited molecules can be released in 10^{-8} sec or less. This is how a laser delivers a very brief and intense flash of light. The geometry that assures an efficient cascade, with quanta inducing transitions instead of "leaking out," also assures that the light of the laser flash is coherent in phase and can form an exceedingly tight (parallel) beam.

The probability of an upward transition is expressed by the integrated area of the corresponding absorption band, in a plot of the extinction coefficient ϵ vs. frequency or wave number (Figure 5.12). For a spontaneous downward transition the probability can be expressed by a rate constant. Imagine a collection of molecules in which N have been excited by a brief flash of light. Assume that induced de-excitation and radiationless processes (photochemistry or conversion to heat) are negligible, so that fluorescence is the only mechanism for de-excitation. The excited population then decays according to

$$dN/dt = -k_f N \tag{5.3}$$

where k_f is the probability per unit time that an excited molecule will return to the ground state by fluorescence. Integration gives

$$N = N_0 e^{-k_f t} = N_0 e^{-t/\tau_0} \tag{5.4}$$

where N_0 is the initial number of excited molecules and $\tau_0 = 1/k_f$. The rate constant k_f is expressed in sec^{-1}; its reciprocal τ_0 is called the intrinsic life-

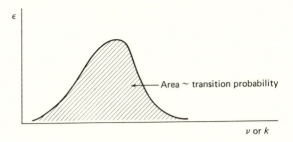

Fig. 5.12. Plot of ϵ vs. ν or k showing probability of an upward transition. (See text for discussion.)

time. When fluorescence is the only avenue for de-excitation, τ_0 is the time needed for the excited population to decline to $1/e$ of the initial number. Since the probability of a spontaneous downward transition is proportional to that of an upward transition, we can write

$$k_f = 1/\tau_0 = \text{const.} \times \int \epsilon \, dk \qquad (5.5)$$

where k is the wave number, $1/\lambda$, in cm^{-1}, and the integration is carried out over the appropriate band in the absorption spectrum. Thus the intrinsic lifetime of a molecule in the excited state varies inversely as the integrated area of the absorption band. The correct formulation of Equation 5.5, for a single electron transition and the corresponding absorption band, is

$$1/\tau_0 = 2.9 \times 10^{-9} \, (1/k_{av}^{-3}) \int (\epsilon/k) \, dk \qquad (5.6)$$

where k is the wave number, ϵ is the extinction coefficient, M^{-1} cm^{-1}, and k_{av}^{-3} is an average of k^{-3} weighted over the fluorescence band:

$$k_{av}^{-3} = \left(\int k^{-3} I_f \, dk \right) \Big/ \left(\int I_f \, dk \right)$$

where I_f is the intensity of fluorescence in quanta per unit interval of wave number. A fair approximation is given by

$$1/\tau_0 = 3 \times 10^{-9} \, k_m^2 \, \Delta k \epsilon_m \qquad (5.7)$$

where k_m and ϵ_m are the wave number and the extinction coefficient (the latter in units of M^{-1} cm^{-1}) at the maximum of absorption, and Δk is the band width at half-maximum. For chlorophylls the lowest excited singlet state (corresponding to the long wave absorption maximum) has an intrinsic lifetime in the range 10 to 40 nsec (1 nsec = 10^{-9} sec).

The actual lifetime of a molecule in an excited state is less than the intrinsic lifetime because mechanisms other than fluorescence contribute to the de-excitation. If we write rate constants (sec^{-1}) k_d for radiationless de-excitation[5] and k_p for photochemical utilization, Equation 5.3 becomes

$$dN/dt = -(k_f + k_d + k_p)N \qquad (5.8)$$

and Equation 5.4 becomes

$$N = N_0 e^{-t/\tau} \qquad (5.9)$$

where the *actual* lifetime τ equals $1/(k_f + k_d + k_p)$. The quantum yield of fluorescence, ϕ_f (quanta emitted $\div$ quanta absorbed), equals the proportion of excited molecules that decay by the fluorescent route:

$$\phi_f = k_f/(k_f + k_d + k_p) \qquad (5.10)$$

With $\tau_0 = 1/k_f$ and $\tau = 1/(k_f + k_d + k_p)$, this equation gives

$$\tau = \tau_0 \phi_f \qquad (5.11)$$

The lifetime is attenuated in proportion to the yield of fluorescence.

Until recently it was difficult to measure these lifetimes directly. But the intrinsic lifetime could be computed from the absorption spectrum by Equation 5.6 or 5.7, and the actual lifetime could be computed from Equation 5.11 if the quantum yield of fluorescence could be measured. For chlorophylls in solution the actual lifetimes measured directly agree well with those computed indirectly. For Chl a, $\tau_0 = 15$ nsec, $\phi_f = 33\%$, and $\tau = 5$ nsec. The value of τ_0 corresponds to $k_f = 6.7 \times 10^7$ sec^{-1}. From the value of τ, $k_f + k_d + k_p = 1/\tau = 20 \times 10^7$ sec^{-1} and with $k_p = 0$ (no photochemistry), $k_d = (20 - 6.7) \times 10^7 = 13.3 \times 10^7$ sec^{-1}.

For a photochemical system we can draw a simple relationship between the quantum efficiency of the photochemistry and the yield of fluorescence. When the system is active, the fluorescence yield is given by Equation 5.10, and the photochemical efficiency is given by

$$\phi_p = k_p/(k_f + k_d + k_p) \qquad (5.12)$$

Now suppose that we can inhibit (poison) the photochemistry so as to reduce k_p to zero while not affecting k_f or k_d. The fluorescence yield then rises to a maximum value:

$$\phi_f(\text{max}) = k_f/(k_f + k_d) \qquad (5.13)$$

By combining Equations 5.10, 5.12, and 5.13 (left as an exercise for the student), we can show that the quantum yield of photochemistry is

$$\phi_p = 1 - \phi_f/\phi_f(\text{max}) = 1 - f/f_{\text{max}} \qquad (5.14)$$

where f and f_{max} are actual fluorescence measurements, uncalibrated but taken under the same conditions. In this way one can estimate the quantum efficiency of the photochemical reaction simply by taking the ratio of two fluorescence measurements. If the photochemistry is 98% efficient, the fluorescence should rise 50-fold when the photochemistry is turned off. The same considerations apply to an ensemble of molecules such as an antenna of Chl with a photochemical reaction center. One must only remember to interpret k_p properly. The effective rate constant for trapping is diminished, in the ensemble, because a quantum spends only a small fraction of its time visiting a reaction center. The effective k_p is less than that in purified reaction centers, by a factor roughly equal to the ratio of antenna molecules to reaction-center pigments. The diminution of k_p is less severe if there is focusing by means of an energetically downhill cascade to the reaction centers.

5.3 Fluorescence in reaction centers and in photosynthetic membranes, in relation to energy transfer and photochemistry

In reaction centers from a photosynthetic bacterium such as *Rhodopseudomonas sphaeroides*, the quantum yield of fluorescence in the 900 nm band (corresponding to the 870 nm absorption band) is about 3×10^{-4} when the reaction centers are in the photochemically active state PIQ. If the quinone is reduced, either chemically (using sodium dithionite, $Na_2S_2O_4$) or by illumination in the presence of an electron donor (Reaction 5.2), the reaction centers enter the state PIQ⁻, and the 900 nm fluorescence becomes three times more intense; the yield rises to 10^{-3}. These facts were known before Bpheo had been implicated as an intermediary electron carrier, and it was supposed that the reduction of UQ blocks the photochemical pathway, bringing the rate constant k_p to zero and allowing an increased fluorescence as expected from Equation 5.10. The quantum efficiency of the photochemistry was then predicted, from Equation 5.14, to be $\phi_p = 1 - \frac{1}{3} = 0.7$. But a direct measurement of the quantum efficiency, from the rate of bleaching of the 870 nm absorption band (see Equation 4.18), gave $\phi_p = 1.02 \pm 0.04$. The quantum efficiency was at least 98%, and Equation 5.14 appeared not be valid.

We now believe that the earliest photochemical step, the one with rate constant k_p, is the transfer of an electron from Bchl (P) to Bpheo (I). Reducing the quinone does not block this step; its aborts the normal transfer to UQ, and the electron returns from I⁻ to P⁺:

$$\text{PIQ}^- \xrightarrow{\ h\nu\ } \text{P*IQ}^- \xrightarrow{\ k_p\ } \text{P}^+\text{I}^-\text{Q}^- \tag{5.15}$$

The return from P⁺I⁻Q⁻ to PIQ⁻ goes partly by way of a triplet state of P870, with subsequent decay to the ground state. In this model we can attribute the threefold rise in fluorescence to a threefold decrease of k_p, caused by the reduction of Q. Such a change of k_p could arise if the transfer of an electron from P to I is retarded by the negative charge on Q^-. An alternative explanation is that the increased fluorescence comes from a back-reaction from $P^+I^-Q^-$ to P^*IQ^-. This back-reaction competes with the main pathway from $P^+I^-Q^-$ to PIQ^- shown in Reaction 5.15. Fluorescence from P^* has the same spectrum, whether the P^* is created in the initial act of light absorption $(P \xrightarrow{h\nu} P^*)$ or by a back-reaction. The former is short-lived, decaying with the rate constant $k_f + k_d + k_p$ (Equation 5.8); it is called the prompt fluorescence. The latter, the delayed fluorescence, persists as long as it is being fed by the back-reaction; its lifetime is essentially that of the state $P^+I^-Q^-$. The lifetime of $P^+I^-Q^-$, measured by the decay of characteristic absorbance changes (see Section 8.1), is about 10^{-8} sec.

In fact, the observed intensities and lifetimes of prompt and delayed fluorescence from these reaction centers do not support either of the foregoing explanations. Most of the rise in total fluorescence caused by reducing the quinone is due to one or more components of delayed fluorescence. However, the various recent reports of delayed fluorescence from reaction centers are in conflict with each other and cannot be correlated convincingly with the lifetimes of known intermediates in the photochemical sequence. At this writing we can only say that the total fluorescence does rise when the quinone is reduced, and this rise remains a convenient indicator of the redox state of the reaction center.

For active reaction centers (state PIQ) there is negligible delayed fluorescence; we can proceed sensibly to compute relations between yield and lifetime of the prompt fluorescence. The intrinsic lifetime τ_0 of the prompt fluorescence of reaction centers has been computed from the area of the 870 nm absorption band (Equation 5.6). The outcome of this computation depends on how one deals with the fact that the 870 nm band is due not to a single molecule but to two or more strongly interacting Bchl molecules in the reaction center. If this complication is ignored, the computed value of τ_0 is 2×10^{-8} sec. Then with a fluorescence yield of 3×10^{-4} in active reaction centers, Equation 5.11 gives $\tau = 6 \times 10^{-12}$ sec or 6 psec for the lifetime of the excited state P^*. These values of τ_0 and τ correspond to reciprocals $k_f = 1/\tau_0 = 3 \times 10^7$ sec^{-1} and $(k_f + k_d + k_p) = 1/\tau = 1.7 \times 10^{11}$ sec^{-1}.

We do not know the value of k_d in reaction centers; it must be small compared with k_p in view of the high photochemical efficiency. Then $k_p \gg k_f + k_d$, and $k_p \approx 1.7 \times 10^{11}$ sec^{-1} in active reaction centers. With the quinone reduced, the threefold greater fluorescence corresponds to an "effective" k_p three times smaller, about 5×10^{10} sec^{-1}.

The most recent direct measurements of the first photochemical step, based on absorbance changes induced by a laser flash, gave a rise time of only 3 psec, corresponding to $k_p \approx 3 \times 10^{11}$ sec^{-1}. This agrees fairly well with the value computed from the yield of fluorescence, in view of the uncertainty involved in computing τ_0.

The values of k_p corresponding to the yields of fluorescence in "active" (PIQ) and "inactive" (PIQ⁻) reaction centers are consistent with measurements of the fluorescence of antenna Bchl in chromatophores. For *Rp. sphaeroides*, with about 30 molecules of antenna Bchl per reaction center, the yield of fluorescence (from the antenna) is about 0.01 when the reaction centers are photochemically active. From Equation 5.11 this corresponds to $\tau = 2 \times 10^{-10}$ sec or 200 psec, and $k_f + k_d + k_p = 1/\tau = 5 \times 10^9$ sec⁻¹. With $k_f + k_d \ll k_p$, as befits a high photochemical efficiency in chromatophores, k_p is close to 5×10^9. The predicted lifetime of 200 psec, for antenna Bchl in its singlet excited state, has been confirmed by direct observation of the decay of fluorescence after a flash. The value of k_p is about one-thirtieth the value in isolated reaction centers; this is compatible with the fact that a quantum of excitation spends most of its time in the antenna and not in the reaction center (see the last paragraph of Section 5.2).

If chromatophores are treated with a reductant so as to reduce the primary quinone of the reaction centers, or illuminated with an electron donor (Reaction 5.2) to reduce this quinone, the yield of fluorescence from antenna Bchl increases to about 0.03. From Equation 5.10, this could result if k_p changes from 5×10^9 to 1.7×10^9. This threefold change of the effective k_p in chromatophores agrees well with the corresponding change in isolated reaction centers.

The fluorescence of antenna Bchl rises by the same factor of three, to 0.03, when chromatophores are treated with an oxidant or illuminated so as to oxidize the special pair of Bchl in the reaction center, $P \longrightarrow P^+$. This suggests that the effective k_p *in the reaction centers* is changed, upon oxidation, from 1.7×10^{11} to 5×10^{10}, just as by reduction of the quinone. But with P870 oxidized, the photochemistry known to us cannot happen. What, then, do we mean by a k_p of 3×10^{10} in these reaction centers? Apparently with the oxidation of P870, a new mechanism arises for quenching the excitation quantum. The new mechanism has a rate constant k_q equal to 5×10^{10} sec⁻¹. Instead of saying that k_p declines from 1.7×10^{11} to 5×10^{10} sec⁻¹, we can say that a k_p of 1.7×10^{11} sec⁻¹ is replaced by a k_q of 5×10^{10} sec⁻¹ when P870 is converted to P^+. From the point of view of the antenna it is all the same; the effective k_p in the chromatophore declines from 5×10^9 to 1.7×10^9 sec⁻¹.

In a mutant of *Rp. sphaeroides* that has no reaction centers, the yield of antenna Bchl fluorescence is 0.10, ten times higher than the yield with active reaction centers.

An antenna pigment–protein complex, free from reaction centers, can be isolated from *Rp. sphaeroides*. When it is mixed with purified reaction centers, the function of efficient energy transfer from antenna to reaction centers is reconstituted. The yields of fluorescence can be made to resemble those in chromatophores, but the agreement is fortuitous because the yields with and without reaction centers can be set by adjusting the proportion of reaction centers to antenna and the concentration of the detergent used in keeping the system soluble.

Measurements of fluorescence and photochemistry allow us to distinguish between two models for the architecture of the photosynthetic membrane, called "puddle" and "lake," mentioned in Section 2.2. In the former each reaction center has its puddle of antenna, in a package (a photosynthetic unit) isolated from its neighbors with respect to energy transfer. In the lake model there are no such barriers to energy transfer; the reaction centers are embedded periodically in a large and continuous array of antenna. We can formulate the basis for making the distinction; in doing so we shall assume for simplicity that k_p (and its counterpart k_q) is zero in inactive reaction centers. In the lake model the fluorescence yield is given by $\phi_f = k_f/(k_f + k_d + k_p)$, and k_p is proportional to the fraction of reaction centers that are active. If this fraction is x, then $k_p = kx$ where k is a constant. Then the reciprocal of ϕ_f is a linear function of the proportion x of active reaction centers.

$$1/\phi_f = 1 + k_d/k_f + kx/k_f \qquad (5.16)$$

In the puddle model the photosynthetic units fall into two categories. In a fraction x, each unit has an active reaction center and a fluorescence yield ϕ_f^a given by Equation 5.10. In the remainder, $1 - x$, the reaction centers are inactive, and the greater yield ϕ_f^i is given by Equation 5.13. The average yield is

$$\phi_f = x\phi_f^a + (1 - x)\,\phi_f^i = \phi_f^i - (\phi_f^i - \phi_f^a)\,x \qquad (5.17)$$

In this case ϕ_f, rather than its reciprocal, varies linearly with x. Similar considerations apply to the photochemical efficiency ϕ_p as given by Equation 5.12. The reasoning remains the same if k_p for inactive reaction centers is not zero, or is substituted by a different rate constant k_q.

Experimental tests of these predictions have been made with various species of photosynthetic bacteria, by measuring either the fluorescence or the photochemical efficiency when progressively greater fractions of the reaction centers are made inactive. This is usually done kinetically, as illustrated in Fig. 5.13. A suspension of cells or chromatophores is exposed to constant illumination, and the intensity of fluorescence is monitored as the sample responds to the light. In a companion experiment the absorbance at 870 nm can be measured to give the rate and extent of P870 oxidation in the reaction centers. The two sets of data can then be compared with reference to Equations 5.16 and 5.17. For the most part these experiments have supported a lake model in photosynthetic bacteria. The puddle model finds some applicability in green plant chloroplasts.

In summary, we can generate a consistent picture of fluorescence and photochemical efficiency in reaction centers and in the chromatophores that

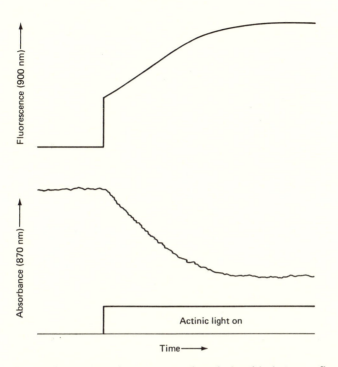

Fig. 5.13. An experiment to test the relationship between fluorescence of antenna Bchl and photo-oxidation of Bchl in the reaction centers, in a suspension of chromatophores from a photosynthetic bacterium such as *Rp. sphaeroides*. The sample is exposed to constant strong actinic illumination, starting at time zero. The intensity of fluorescence (elicited by the actinic light) is monitored at 900 nm, and the absorbance at 870 nm is monitored by means of a weak beam of 870 nm light. Loss of 870 nm absorbance signified oxidation of Bchl (P870) in the reaction centers, and the rate of this absorbance decrease gives a relative measure of the quantum efficiency of this photochemical process.

contain them, but we do not understand the basis of the increased fluorescence that comes when the reaction centers are driven to inactive states (PIQ$^-$, P$^+$IQ$^-$, and P$^+$IQ). The reaction centers appear to be embedded in an ocean of antenna pigment.

5.4 Antenna pigment–protein components: varieties, association with reaction centers, and physiological regulation

Photosynthetic membranes of bacteria show great variety in their absorption spectra in the near infrared, where nearly all the absorption is due to Bchl. In *Rhodospirillum rubrum* there is just one major band at 885 nm, due to antenna Bchl *a. Rhodopseudomonas viridis* shows a single band at 1020 nm (the long wave record for an antenna pigment), due to Bchl *b*. In wild type strains of *Rp. sphaeroides* the antenna Bchl *a* shows bands at 800, 850, and 875 nm (see Fig. 2.11), but in carotenoidless mutants of this species there is only one major band, near 860 nm (Fig. 5.8). Other species with Bchl *a* show bands at 800, 820, 850, and 890 nm, all in the same organism. All of these bands of Bchl *a*, whose diversity is attributable to interactions between Bchl and its surroundings, can be associated with just three types of antenna Bchl–protein component. First there is B875 (also called B890), a component in which the Bchl *a* shows a single major band somewhere in the range 870-890 nm. In *Rp. sphaeroides* there are about 30 molecules of Bchl as B875 for every reaction center. In carotenoidless mutants the single band at 860-870 nm seems analogous to B875; it also occurs in fixed proportion to the reaction centers. This does not mean that the synthesis of B875 is linked to the formation of reaction centers; some mutants that lack reaction centers are capable of forming B875. No mutant has been found, however, that has reaction centers and lacks B875.

The second type of antenna component has two absorption bands, one near 800 nm and another near 850 nm. This component has been purified and characterized extensively; it is called B850. The ratio of B850 to B875 in a photosynthetic bacterium varies with the conditions of growth (see below in this section). A third component, B820, has bands near 800 and 820 nm. It is closely similar to B850 and might be a minor variant of the latter. B820 isolated from *Chromatium vinosum* can apparently be converted to B850, and back again to B820, by adding and removing specific detergents.

When challenged to grow in very weak light, photosynthetic bacteria respond by making more antenna pigment and more reaction centers per cell. This behavior was first described quantitatively by G. Cohen-Bazire, W. R. Sistrom, and R. Y. Stanier in 1957 (see the suggested reading for this chapter). They maintained continuous cultures of *Rp. sphaeroides*, exposed to various light intensities and flushed with either nitrogen or air. Changes of cell mass and pigment content were determined for cultures subjected to an abrupt change of light intensity or atmosphere. The results of such experiments are outlined in Fig. 5.14, where the logarithms of cell mass and Bchl content are plotted against time. In balanced exponential growth these plots are parallel straight lines. When such a culture, growing under N_2, is subjected to an increase of light intensity, growth (measured as cell mass) continues at approximately the same rate, but the synthesis of Bchl is suspended. As growth continues in strong light, the Bchl per cell diminishes. Eventually Bchl synthesis is resumed, and the culture reaches a new condition of balanced growth at a lower density of pigments. If the light intensity is then returned to its original weak level, growth slackens because there is too little pigment to sustain the former near-maximal rate. Concurrently an accelerated synthesis of Bchl restores the original density of pigmentation, and the origi-

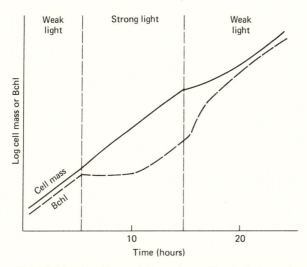

Fig. 5.14. A culture of *Rp. sphaeroides*, in balanced exponential growth under nitrogen in weak light, is transferred abruptly to strong light. Growth continues and Bchl synthesis stops until the density of pigmentation reaches a lower level, when balanced growth is resumed. If the culture is returned to weak light, growth slackens and Bchl synthesis is accelerated until the original density of pigmentation has been restored. Changing from N_2 to air has the same effect as raising the light intensity.

nal condition of balanced growth is attained again. Substitution of O_2 (or air) for N_2 has the same effect as an increase of light intensity. The strongest suppression of pigmentation is achieved with a combination of light and O_2. Cultures growing in the dark make less pigment per unit cell mass under vigorous aeration and become more densely pigmented when the supply of O_2 is restricted. It is for this reason that colonies growing on culture plates, aerobically in darkness, are very pale when they are small and acquire more color when their size limits the diffusion of O_2 to their interiors.

This regulation of pigment synthesis is quantitatively spectacular and nicely tuned: Under anaerobic conditions in the light, the density of pigmentation reaches a level just great enough to afford a near-maximal rate of growth at the prevailing light intensity. This can be seen in Fig. 5.14, where a culture in balanced growth in weak light shows little change of growth rate when the light is made stronger.

In an organism such as *Rs. rubrum*, in which B875 is the sole antenna component, the ratio of B875 to reaction centers remains constant throughout these changes. The ratio of B875 to intracytoplasmic (chromatophore) membrane protein also remains constant; denser pigmentation is brought about by making more intracytoplasmic membrane and not by increasing the concentrations of pigments in these membranes. In these organisms the response is relatively crude, in that the cells may overproduce intracytoplasmic membrane and reaction centers while making more antenna pigment in response to weak light. A more sophisticated system is found in bacteria, such as wild type *Rp. sphaeroides*, which have B850 as well as B875 and reactions centers. Such bacteria show two responses to a decrease of light intensity during growth. First, the ratio of B875 (and reaction centers, and intracytoplasmic membrane) to cell mass increases. Second, the ratio of B850 to B875 increases. The intracytoplasmic membranes themselves become more densely pigmented; they acquire increasing proportions of B850 in addition to their fixed proportions of B875 and reaction centers. In a change from moderately bright to very weak light the content of B875 and reaction centers (per unit cell mass) may change 2.5-fold, whereas the total Bchl content changes 25-fold owing to the extra B850. More antenna pigment serves each reaction center – about 300 Bchl instead of 30. At the other extreme, cultures growing in the light under vigorous aeration become nearly depleted of intracytoplasmic membranes and pigments, but these "bleached" cells retain their ability to form pigments under appropriate conditions of growth.

The mechanism(s) of this striking example of cellular differentiation remains obscure. Increased pigmentation goes hand in hand with the synthesis of new intracytoplasmic membrane protein. When protein synthesis is poisoned, the entire response is suppressed, and there is no significant accumulation of biosynthetic precursors of Bchl. In the biosynthesis of Bchl *a* (Fig. 2.7) the rate-limiting step appears to be the formation of δ-aminolevulinic acid from succinic acid and glycine, mediated by the enzyme ALA synthase (if δ-aminolevulinic acid is fed to a culture, so as to bypass the limitation, the rate of Bchl synthesis rises). ALA synthase has been purified. It is inhibited by ATP, which should be abundant in cells growing under strong light and/or aeration. It is activated (converted from an inactive to an active form) by trisulfides such as cystine trisulfide, and the active form of the enzyme predominates in aerated cultures. ALA synthase is also inhibited by heme. These facts are points of attack on a problem that may have a multiple solution.

B850 is a complex in which three molecules of Bchl *a* and one carotenoid molecule are bound (noncovalently) to a protein composed of two polypeptides, each weighing about 8500 daltons. This complex, purified from chromatophores by a combination of differential solubility in detergents and differential centrifugation, has the same absorption spectrum in vitro as in the living cell. B875 has been isolated from *Chr. vinosum* and from *Rs. rubrum;* it contains two Bchl *a* and one carotenoid per about 20 kdalton of protein. The recently isolated B875 of *Rp. sphaeroides* is similar except that it has *two* carotenoid molecules associated with two Bchl *a* and about 20 kdalton of protein (two polypeptides). B860, the apparent counterpart of B875 in carotenoidless mutant *Rp. sphaeroides*, has also been purified; it also has two Bchl *a* (but no carotenoids) associated with two polypeptides similar to those of B850. Conceivably the most important difference between B850 and B875 is that the former has a binding site for one more Bchl (per unit complex) than the latter, with the extra Bchl accounting for the absorption band at 800 nm in B850.

The absence of B850 in carotenoidless mutants of *Rp. sphaeroides* is a genetic curiosity. The mutation of a single genetic locus[6] leads to failure of a step in carotenoid biosynthesis, and to an inability to form B850.

SUGGESTED READINGS

Duysens, L. N. M., Huiskamp, W. J., Vos, J. J., and van der Hart, J. M. (1956). Reversible changes in bacteriochlorophyll in photosynthetic bacteria upon illumination. *Biochim. Biophys. Acta 19*, 188–90.

Goedheer, J. C. (1959). Reversible oxidation of pigments in bacterial chromatophores. *Brookhaven Symp. Biol. 11*, 325–31.

Clayton, R. K. (1963). Toward the isolation of a photochemical reaction center in *Rhodopseudomonas sphaeroides*. *Biochim. Biophys. Acta 75*, 312–23.

Clayton, R. K., and Wang, R. T. (1971). Photochemical reaction centers from *Rhodopseudomonas sphaeroides*. *Methods Enzymol. 23*, 696–704.

Parson, W. W. (1968). The role of P_{870} in bacterial photosynthesis. *Biochim. Biophys. Acta 153*, 248–59.

Sistrom, W. R., and Clayton, R. K. (1964). Studies on a mutant of *Rhodopseudomonas sphaeroides* unable to grow photosynthetically. *Biochim. Biophys. Acta 88*, 61–73.

Straley, S. C., Parson, W. W., Mauzerall, D. C., and Clayton, R. K. (1973). Pigment content and molar extinction coefficients of photochemical reaction centers from *Rhodopseudomonas sphaeroides*. *Biochim. Biophys. Acta 305*, 597–609.

Feher, G., and Okamura, M. Y. (1977). Reaction centers from *Rhodopseudomonas sphaeroides*. *Brookhaven Symp. Biol. 28*, 183–93.

Strickler, S. J., and Berg, R. A. (1962). Relation between absorption intensity and fluorescence lifetime of molecules. *J. Chem. Physics 37*, 814–22.

Zankel, K. L., Reed, D. W., and Clayton, R. K. (1968). Fluorescence and photochemical quenching in photosynthetic reaction centers. *Proc. Natl. Acad. Sci. U.S. 61*, 1243–9.

Vredenberg, W. J., and Amesz, J. (1967). Absorption characteristics of bacteriochlorophyll types in purple bacteria and efficiency of energy transfer between them. *Brookhaven Symp. Biol. 19*, 49–61.

Campillo, A. J., Hyer, R. C., Monger, T. G., Parson, W. W., and Shapiro, S. L. (1977). Light collecting and harvesting processes in bacterial photosynthesis investigated on a picosecond time scale. *Proc. Natl. Acad. Sci. U.S. 74*, 1997–2001.

Vredenberg, W. J., and Duysens, L. N. M. (1963). Transfer of energy from bacteriochlorophyll to a reaction centre during bacterial photosynthesis. *Nature (London) 197*, 355–7.

Clayton, R. K. (1967). An analysis of the relationships between fluorescence and photochemistry during photosynthesis. *J. Theor. Biol. 14*, 173–86.

Heathcote, P., and Clayton, R. K. (1977). Reconstituted energy transfer from antenna pigment–protein to reaction centres isolated from *Rhodopseudomonas sphaeroides*. *Biochim. Biophys. Acta 459*, 506–15.

Thornber, J. P., Trosper, T. L., and Strouse, C. E. (1978). Bacteriochlorophyll in vivo: Relationship of spectral forms to specific membrane components. In *The Photosynthetic Bacteria*, R. K. Clayton and W. R. Sistrom, eds., Chapter 7, pp. 133–60. Plenum, New York.

Cohen-Bazire, G., Sistrom, W. R., and Stanier, R. Y. (1957). Kinetic studies of pigment synthesis by non-sulfur purple bacteria. *J. Cell. Comp. Physiol. 49*, 25–68.

Aagard, J., and Sistrom, W. R. (1972). Control of synthesis of reaction center bacteriochlorophyll in photosynthetic bacteria. *Photochem. Photobiol. 15*, 209–25.

Lascelles, J. (1978). Regulation of pyrrole synthesis. In *The Photosynthetic Bacteria*, R. K. Clayton and W. R. Sistrom, eds., Chapter 42, pp. 795–808. Plenum, New York.

Clayton, R. K., and Clayton, B. J. (1972). Relations between pigments and proteins in photosynthetic membranes of *Rhodopseudomonas sphaeroides*. *Biochim. Biophys. Acta 283*, 492–504.

Cogdell, R. J., and Thornber, J. P. (1979). The preparation of characterization of different types of light-harvesting pigment–protein complexes from some purple bacteria. In *Chlorophyll Organization and Energy Transfer in Photosynthesis (Ciba Found. Symp. No. 61, New Series)*, pp. 61–79. Excerpta Medica, Amsterdam.

Zankel, K. L. (1978). Energy transfer between antenna components and reaction centers. In *The Photosynthetic Bacteria*, R. K. Clayton and W. R. Sistrom, eds., Chapter 18, pp. 341–7). Plenum, New York. (Also see Zankel and Clayton in the suggested readings for Chapter 2.)

Sauer, K. (1978). Photosynthetic membranes. *Acc. Chem. Res. 11*, 257–64.

6 Photosynthetic membranes of plants: components and their molecular organization; energy transfer and its regulation

6.1 Pigment-protein components of chloroplasts; their relation to gross visible structures and to Photosystems 1 and 2

With the discovery that green-plant photosynthesis involves the cooperation of two photosystems, it was natural to ask how each of these systems is served by antenna pigments. This could be explored by measuring action spectra for the enhancement phenomenon, and more specifically by measuring action spectra for partial reactions involving only Photosystem 1 or Photosystem 2, using suspensions of chloroplasts or chloroplast fragments.[1] The operation of each photosystem can be studied by itself through the use of suitable reagents, mainly electron donors or acceptors and specific inhibitors of electron transport (see Part III). Mutants can also be used, but they may be abnormal in their pigmentation as well as in their photochemical activities.

The result of such measurements, in general terms, has been that Photosystem 1 is driven preferentially by a "long wave" component of antenna Chl a, with appreciable absorption at wavelengths greater than 680 nm, whereas Photosystem 2 is driven better by a shorter wave form of Chl a, with little absorption beyond 680 nm. Chl b and phycobilins, when present, are better sensitizers of Photosystem 2, and carotenoids serve both systems. These findings emphasized the question, "How do plants in their natural environment achieve balanced operation of the two photosystems?" This was a difficult and intriguing question, especially because the quantum efficiency of photosynthesis, measured in the green alga *Chlorella*, was known to be nearly constant under monochromatic light ranging from 400 to 680 nm. Over this range of wavelengths there surely were significant variations in relative absorption by the pigments associated primarily with Photosystem 1 and Photosystem 2, respectively. One could readily conceive a "spillover" mechanism, such that excess energy absorbed by the pigments of Photosystem 2 could be diverted to Photosystem 1. Spillover in the reverse direction is unlikely in view of the red drop phenomenon (Section 3.2). But, as we shall see in Section 6.3, a more elaborate regulatory system is closer to reality.

In the 1960s many investigators (notably J. M. Anderson and N. K. Board-

111

man) made intensified efforts to isolate specific components or fractions, identifiable with one or the other photosystem, from chloroplast thylakoid membranes. The general technique was to disrupt thylakoids into fragments, with or without the help of a detergent, and isolate components of different densities by centrifuging the material through a gradient of increasing sucrose concentration.

Moderate disruption yielded a buoyant fraction, Fraction I, enriched for Photosystem 1 activity and devoid of Photosystem 2 activity, and a dense fraction, Fraction II, having both activities but enriched for Photosystem 2. When examined in the electron microscope, Fraction I consisted of small vesicles whose surfaces resembled those of stroma lamellae (see the introduction to Part II), and Fraction II looked like pieces of membrane in the stacked region, the grana lamellae. It appeared that stroma lamellae are far more susceptible to disruption than grana lamellae, and the broken stroma lamellae are more buoyant than the remaining material. The use of a stronger (more disruptive) detergent gave a higher yield of Fraction I and a lower yield (but with better enrichment for Photosystem 2) of Fraction II, as if pieces of Fraction I were being released from Fraction II.

The following tabulation is representative of the composition and activities of Fractions I and II:

Fraction I	*Fraction II*
40 Chl a (long wave form) per 1 P700	Some P700; long and shorter wave forms of Chl a
No Chl b	Chl a/Chl b $<$ 3 (3 is the value for entire chloroplasts)
Photosystem 1 activity; no Photosystem 2 activity	Both photosystems active; relatively more Photosystem 2 activity
May have Cyt f, Cyt 563, and β-carotene, depending on choice of detergent	Cyt 559 present

P700 is the primary photochemical electron donor in Photosystem 1; it is a "special pair" of Chl a in the reaction center. In this respect the reaction center of Photosystem 1 is analogous to that of a photosynthetic bacterium. The cytochromes f, 563, and 559 will be a subject of Part III.

A marked improvement in the characterization of thylakoid membrane components came when Fractions I and II were subjected to further analytical treatments, which included permeation chromatography (a method of sorting molecules and macromolecular entities by their size), chromatography through hydroxylapatite, and electrophoresis in polyacrylamide (see Section 5.1). By these procedures J. P. Thornber and collaborators discovered a major antenna pigment–protein complex that contains roughly equimolar amounts of Chls a and b. This complex, called LHa/b (the LH for light-harvesting), has no photochemical activity. In the intact membrane, LH$a/b$ transfers excitation energy to both photosystems, but predominantly to Photosystem 2. All of the Chl b of chloroplasts is found in LH$a/b$.

In Thornber's procedure for isolating LH$a/b$, the chloroplast fragments were exposed to the detergent sodium dodecyl sulfate (SDS), which abolished Photosystem 2 activity. Aside from LH$a/b$, Thornber found only a fraction like the earlier Fraction I, plus free Chl a amounting to about 30% of the total Chl a of the chloroplasts. Other workers, using milder detergents such as digitonin and Triton X-100, have isolated Chl a-protein complexes with Photosystem 2 activity.[2] These studies suggest that thylakoid membranes contain three principal components: a major antenna component (LH$a/b$) and a component for each photosystem, consisting of a reaction center and a "subantenna" of Chl a. This tripartite organization appears to be universal in green plants and algae, with certain qualifications (see later, Fig. 6.1 and Section 6.4). In blue-green and red algae the phycobilisomes take the place of LH$a/b$, and in some marine algae Chl b is replaced by Chl c.

In describing these components of the photosynthetic membrane I shall use the terms "Core 1" and "Core 2" to denote the hypothetical entities, the reaction center and associated subantenna of each photosystem (plus the supporting framework of protein and phospholipid). In using the concept of Cores 1 and 2 we should keep in mind that there is no morphological entity in the membrane that can be identified decisively with either Core 1 or Core 2, although there are suggestive correlations to be described later in this chapter.

The pigment-protein complexes isolated from green plant tissues by various investigators have received a profusion of names, especially Complex I or II and Fraction I or II. The earlier fractions designated "II," described before LH$a/b$ had been recognized as a separate entity, were generally mixtures of LH$a/b$ and components of Photosystem 2. Henceforth we shall use "Fraction 1" and "Fraction 2" as generic terms for isolated Chl-protein complexes that seem to correspond to all or part of Core 1 and Core 2, respectively. These designations should not be confused with the Fractions I and II discussed earlier.

In chloroplasts having stacked regions (grana), Photosystem 2 activity is confined mainly to fractions derived from these regions.

Fraction 1 contains long wave Chl a (several forms, absorption maxima near 675–695 nm) and P700, the photochemical electron donor of Photosystem 1. These pigments are associated with an assortment of polypeptides, the most conspicuous association being between P700 and a pair of polypeptides weighing 50–70 kdalton each.[3] Lesser polypeptides are associated with other electron carriers in Photosystem 1, and perhaps with the subantenna. C. Bengis and N. Nelson (see the suggested reading) studied the photochemical activities of Photosystem 1 fractions (derived from Swiss chard) from which various polypeptides had been removed selectively, by the use of detergents. They concluded that P700 is bound to a pair of 70-kdalton polypeptides, and that three smaller polypeptides (with associated functional groups) are involved in accepting electrons from P700 (in consequence of the photochemistry) and transferring them to more remote acceptors. Some details of Bengis and Nelson's conclusions are given in Fig. 8.6, in connection with a discussion of the reaction center of Photosystem 1.

In most cases, well-purified examples of Fraction 1 retain about 40 molecules of Chl *a* for each P700. Bengis and Nelson's "depleted" preparation, with certain polypeptides stripped away, retained 10 to 20 Chl *a* per P700. They suggested that this antenna Chl is bound to the same polypeptides that bind P700. Gentler methods of purification usually yield a Fraction 1 with about 110 Chl *a* per P700, and with additional polypeptides. Perhaps 40 Chl *a* are normally bound to the polypeptides that bind P700, and 70 more are bound to other protein components in the subantenna of Photosystem 1.

No pheophytin *a* has been detected in Fraction 1; in this respect the reaction center of Photosystem 1 is not analogous to bacterial reaction centers. Fraction 1 seems to be absent in some mutant plants and algae that lack P700 and have no Photosystem 1 activity, but it may be that the absence of reaction centers has altered the behavior of the remaining parts of the complex, so that they have been missed in the usual analytical procedures.

LH*a/b* contains Chls *a* and *b* in the ratio *a/b* = 1.2 to 1.4, associated with a protein whose major polypeptides weigh 24 to 28 kdalton. About five such polypeptides can be distinguished by electrophoresis; they could be minor variants of each other. The complex contains about half of the total thylakoid protein and about 60% of the total Chl. The ratio of Chls to protein is uncertain; some Chl may be set free in the process of isolation. There appear to be at least three molecules of Chl *a* and three of Chl *b* per 35 kdalton of protein. The major components of LH*a/b* are missing in mutant plants that lack Chl *b*, but some of the minor polypeptides of this complex can be found in such mutants. Most mutants that are deficient in either Photosystem 1 or Photosystem 2 activity have apparently normal LH*a/b*. The Chl *a* in purified LH*a/b* exists in two or three forms with absorption maxima between about 660 and 675 nm; the absorption maximum of Chl *b* is at 650 nm.

Fraction 2, isolated from chloroplasts with the help of detergents, contains Photosystem 2 activity and Chl *a*, but no Chl *b* and no Photosystem 1 activity. The protein of this fraction is composed of two polypeptides weighing 27 and 43 kdalton.[4]

All three of these isolated complexes contain carotenoids. The ratios of Chls to protein are probably less than in the intact membrane because some Chl is released during fractionation. For the entire chloroplast there is one reaction center of each kind (Photosystems 1 and 2) per about 300 molecules of Chl *a* and 100 of Chl *b*.

The results of more recent fractionations, combined with observations of fluorescence from the Chl-protein components (see Section 6.3), have in-

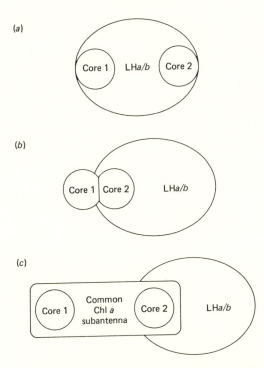

Fig. 6.1. Successive models proposed for the spatial relationships between pigment–protein complexes in the thylakoid membranes of chloroplasts. Core 1 and Core 2 denote the reaction centers of Photosystems 1 and 2, respectively, plus a subantenna of Chl *a* for each. LH*a/b* is an antenna component with Chls *a* and *b* but no reaction centers. Scheme *c* is speculative; there is no evidence for a pool of undifferentiated Chl *a* embracing Core 1 and Core 2.

spired various models for the topological relationships between the components. The earliest picture after the discovery of LH*a/b* was the one shown in Fig. 6.1*a*, with subsequent revision to Fig. 6.1*b*. Figure 6.1*c* is a speculative alternative to Fig. 6.1*b*. Time, it is hoped, will tell what the correct configuration is.

In addition to these integral components of the thylakoid membrane, there are components that appear as particles on the surface and can be washed off. One such component is an enzyme responsible for CO_2 fixation (see Chapter 11), called ribulose diphosphate carboxylase (RuDP carboxylase). This enzyme can be removed by washing the thylakoids with water or buffers of low ionic strength. A second surface component is CF_1, the main part of the enzyme that functions in photosynthesis to form ATP from ADP and phosphoric acid. Component CF_1 is connected to F_0, a hydrophobic protein embedded in the membrane. The complete $CF_1 \cdot F_0$ enzyme can be isolated

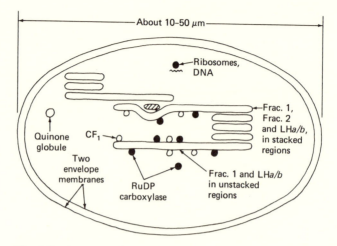

Fig. 6.2. A cartoon (not to scale) of a chloroplast, showing the lamellar thylakoids, the pigment–protein complexes in these thylakoids, and various inclusions and membrane surface particles.

by careful use of detergents, but CF_1 can be released simply by washing
thylakoids with a solution of ethylene diamine tetraacetic acid (EDTA), an
agent that binds divalent cations. Other surface components, not readily seen
but detected by their affinities for specific antibodies, include ferredoxin and
NADP reductase, both of which carry electrons from Photosystem 1 to
$NADP^+$ (see Section 9.2).

Within the chloroplast, but not attached to the thylakoids, one can also
find particles of RuDP carboxylase, starch grains, globules of plastoquinone,
ribosomes, and DNA. The DNA, distinct from that in the cell nuclei, endows
chloroplasts with considerable genetic autonomy.

Some of the foregoing material is recapitulated in Fig. 6.2, a cartoon of a
chloroplast.

6.2   Fine structure of thylakoid membranes as seen with the electron microscope; relation of structural features to functional components

Surface contours of thylakoids, and the contours of fracture planes
between the inner and outer membrane surfaces, have been revealed by electron microscopy using the techniques of freeze-etch and freeze-fracture. The
thylakoids are first frozen rapidly in ice. If superficial ice is sublimed away
(freeze-etch), the outer membrane surfaces are exposed to view. Fracturing
the sample, followed by sublimation, can expose the inner membrane surfaces. Fracture also exposes interfaces within the finite thickness of the

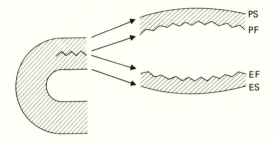

Fig. 6.3. A sketch showing four regions of the thylakoid membrane, including two faces generated by fracture. (See text for details.)

membrane. These interfaces are quite specific; they reflect the paths of least resistance to cleavage, where the material inside the membrane is the most hydrophobic. We thus have access to four specific regions of the thylakoid membrane: inner and outer surfaces, and the two complementary faces of an interface generated by fracture. The nomenclature for these regions has, as usual, proliferated. We shall adopt the terms used by C. J. Arntzen and A. Staehelin, in which the regions inside the chloroplast but outside the thylakoids are called protoplasmic, and the regions confined by the thylakoids are called exoplasmic. The outer surface is then called PS (protoplasmic surface), the fracture face toward the outside is called PF (protoplasmic fracture face), and so forth (Fig. 6.3). Subscripts u and s can be used to denote membranes from unstacked and stacked regions: PS_u, PS_s, etc.

The outer surfaces of PS_u show bumps about 15 nm in diameter, which are due to RuDP carboxylase (about 30%) and CF_1 (about 70%). These bumps can be washed away prior to preparing the specimen for microscopy (see the previous section). They are scarce or absent in the stacked regions (PS_s). Both PS_u and PS_s show other bumps, of about 9 nm diameter, which may be reflections of bodies seen more clearly on the fracture faces. The inner surfaces ES_u have a rather nondescript pebbled aspect. ES_s shows many medium to large bumps on a smooth background; these bumps may be partly reflections of bodies seen in the fracture faces, and partly other components such as the apparatus for O_2 evolution.

The fracture faces show a greater wealth of detail and regularity, especially in the stacked regions. In the unstacked regions one finds many bumps on PF_u, falling roughly into two size groups, about 8 and 11 nm in diameter.[5] As could be expected, EF_u has the complementary appearance of many pits, plus a few 11 nm bumps. In EF_s the dominant feature is a profusion of large bumps, about 16 nm in diameter, plus some 11 nm bumps. Sometimes the 16 nm bumps are seen as four-lobed bodies deployed in a quasicrystalline array (rectangular, with approximately 17.5 × 20 nm spacing), but this is a rare and highly atypical observation. PF_s has large pits complementary to these large bumps, and also some small (8 nm) bumps.

These features can be summarized by saying that the 16 nm bodies are confined to stacked regions, and small (8 and 11 nm) bodies exist in both stacked and unstacked regions. There is great latitude for speculation in identifying these bumps and pits with the isolated pigment–protein components Fraction 1, Fraction 2, and LH$a/b$. Some guides to speculation are: (1) In chloroplasts with stacked thylakoids, both Photosystem 2 activity and the 16 nm bumps are confined to the stacked regions. But in chloroplasts from many plants and algae, with normal Photosystem 2 activity, the thylakoids have no stacked regions and no large bumps. (2) Some mutant plants and algae have no Chl b, no LH$a/b$, and no large bumps (only 8–9 nm bumps), even in their stacked regions. These mutants have Photosystem 2 activity. The same properties can be induced in "normal" bean plants by forcing them to develop under intermittent illumination (2 min light, 2 min dark, etc.). (3) Despite the presence of stacking in the foregoing plants that appear to lack LH$a/b$, other correlations suggest that a component of LH$a/b$ (perhaps a minor polypeptide) is required for stacking. (4) At least in the green alga *Chlamydomonas*, the process of stacking may itself induce the generation of large bodies (16 nm bumps). A mutant of this alga normally has unstacked thylakoids only, and shows fracture faces typical of unstacked thylakoids. It has been possible to induce stacking, and then some of the fracture faces have acquired the appearance typical of stacked thylakoids, with large bumps. On the other hand, in spinach chloroplasts the addition of EDTA causes separation (unstacking) of the thylakoids without loss of the large bumps.

The interpretation currently favored by Arntzen and Staehelin is that the large bodies represent units with Photosystem 1, Photosystem 2, and a full share of LH$a/b$. The 11 nm bumps represent partial units, with Photosystem 1 and/or Photosystem 2 and half of the full share (a subunit?) of LH$a/b$. The 8–9 nm bumps reflect Photosystem 1 and/or Photosystem 2 without LH$a/b$.

These speculations could be aided if the localization of Chls in the membrane could be determined with resolution of 100 Å or better. This might be achieved through photoelectron microscopy. Ultraviolet light can eject electrons from molecules, and the yield of these photoelectrons is exceptionally high from Chls and other tetrapyrroles. If a sample is placed in an electron microscope and exposed to ultraviolet light, the ejected electrons can be focused to the corresponding points in the image plane. A latex sphere coated on one side with a monolayer of Chl glows like a half moon when viewed in this way.

Many observations testify to a directionality in the structure of the thylakoid membrane, such that the evolution of O_2 occurs within the thylakoids and the reduction of NADP happens on the outside, in the protoplasmic space. We shall consider this vectorial property more thoroughly in later chapters, but a few details can be listed now: Most of the enzymes and coenzymes that mediate the reduction and utilization of NADP and the assimilation of CO_2 are on the outside, either on the membrane surface or in the protoplasm; they include ferredoxin, NADP reductase, NADP itself, and RuDP carboxylase. Conversely, the liberation of O_2 and H^+ from H_2O happens inside the thylakoids. A pool of plastoquinone that mediates the flow of electrons from Photosystem 2 to Photosystem 1 has access to both sides of the membrane. When plastoquinone is reduced by electrons from Photosystem 2, H^+ is taken up from outside the thylakoid ($PQ^- \cdot + H^+ \longrightarrow PQH$ and $PQ^{2-} + 2H^+ \longrightarrow PQH_2$), whereas the reoxidation of PQH_2 (with electrons going to Photosystem 1) is attended by the release of H^+ inside the thylakoids. Thus light causes the transfer of protons into the thylakoids from the external protoplasm, a prelude to ATP formation.

6.3 Fluorescence and energy transfer in the thylakoid membrane, and the control of quantum distribution to the two photosystems

The problem of optimal balance in the rates of operation of Photosystems 1 and 2 seems to have its solution in a regulatory mechanism whereby the partitioning of quantum energy between the two photosystems depends on the state of the chloroplast. Quanta absorbed by LH$a/b$ may go to one photosystem or the other, in varying proportions but always mainly to Photosystem 2. Varying numbers of quanta may "escape" from the sub-antenna of Photosystem 2 and be captured by Photosystem 1 (spillover). These changes in quantum distribution are associated with the wavelength of previous illumination, with changing concentrations of divalent cations (especially Mg^{2+}), and with visible movements of the thylakoids (swelling, shrinking, stacking, and unstacking). To complete the network of interactions, light causes Mg^{2+} to move out of the thylakoids (reflexive to inward H^+ pumping), and divalent cations have a dramatic effect in promoting stacking.

One can imagine that imbalanced operation of the two photosystems leads to redistribution of Mg^{2+}, which causes movements of the thylakoids. The pigment–protein complexes might then undergo changes in their respective proximities, causing changes in the rates of transfer of excitation energy among them. The resulting changes of quantum distribution could tend to correct the imbalance in operation of the photosystems. To explore this putative mechanism let us examine the known relationships between the various factors. A major tool in developing this knowledge has been the observation of fluorescence, indicative of excitation energy in the pigmented components.

At room temperature the fluorescence of chloroplasts is confined almost entirely to a single band centered at 685 nm. The more rapid light-induced changes in the yield of this fluorescence can be correlated with changes in the state of the reaction center of Photosystem 2, and not with changes in Photosystem 1. Under physiological conditions the fluorescence rises by a

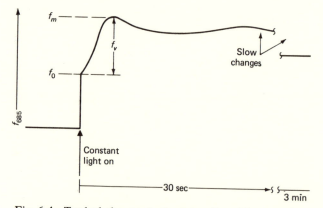

Fig. 6.4. Typical changes in the intensity of fluorescence from Chl a in green plant tissues, during constant illumination after a period of dark adaptation. The early change f_v is correlated with the reduction of the electron acceptor (plastoquinone) in the reaction center of Photosystem 2. The subsequent slow changes reflect variations in the distribution of quanta to the two photosystems. The time scale depends on the light intensity; the early changes are more rapid in stronger light.

factor of 2 to 3 when the electron acceptor of Photosystem 2, a molecule of plastoquinone, is reduced. Because of the close coupling between the 685 nm fluorescence and the state of the Photosystem 2 reaction center, this fluorescence probably arises from Core 2.

The evolution of this fluorescence during constant illumination is portrayed in Fig. 6.4. The value f_0 is established at the onset of illumination. It shows that even when the reaction centers are fully active, some energy is lost as fluorescence. The fluorescence rises to f_m (the difference $f_m - f_0$ is called f_v, for variable fluorescence) as the reaction centers of Photosystem 2 are driven from the state $P \cdot Q$ to $P \cdot Q^-$. Here P represents the photochemical electron donor, a special pair of Chl a molecules. The oxidized species P^+ does not accumulate appreciably under physiological conditions because it is re-reduced rapidly by electrons from water.

After the rapid rise from f_0 to f_m, the fluorescence exhibits slower changes over the course of minutes. These slow changes were eventually ascribed (by C. Bonaventura and J. Myers) to alterations in the distribution of quanta to the two photosystems. Greater fluorescence reflects a greater share of quanta going to Photosystem 2. Next it was recognized that far red illumination (used preferentially by Photosystem 1) causes more quanta to be sent to Photosystem 2, giving more fluorescence, and shorter wave illumination (especially light absorbed by phycobilins in the case of red or blue-green algae) has the reverse effect. These interpretations of fluorescence have been substantiated by measurements of quantum efficiency; far red preillumination raises the efficiency of Photosystem 2 by causing a greater share of absorbed quanta to reach that photosystem, and vice versa.

The state of higher fluorescence induced by far red light was called State I, and the low-fluorescence state was termed State II. If chloroplasts are driven into State I or II, then allowed to rest in the dark, and then given constant illumination, the early change of fluorescence is exhibited as usual, but the magnitudes of f_0 and f_v depend on the state into which the chloroplasts have been put. Both f_0 and f_v are greater in State I than in State II.

In chloroplast fragments the fluorescence properties of States I and II can be mimicked by the addition or withholding of cations, especially the divalent Ca^{2+} and Mg^{2+}. N. Murata and others have shown that in a low-salt medium the fluorescence is low, as in State II. Addition of about 3 mM Ca^{2+} or Mg^{2+}, or about 100 mM Na^+, raises the fluorescence to a level equal to or greater than that of the light-induced State I. The connection between light and cations, in affecting the fluorescence, could lie in the fact that light induces the movement of Mg^{2+} out of the thylakoids. This is a passive efflux in response to the active inward pumping of H^+ that is coupled to the light-initiated electron transport; there is also a passive influx of Cl^-. Measurements of light-induced Mg^{2+} efflux suggest that the concentration of Mg^{2+} could change as much as 10 mM around the thylakoids. It is not clear, however, how the efflux of Mg^{2+} should be related selectively to the long wave light that induces State I, nor is it clear whether the internal or the external Mg^{2+} concentration dictates the distribution of quanta. At present any connection between divalent cations and the light-generated States I and II must be regarded as tenuous.

These effects of light and cations are probably related in some way to the specific effects of cations on the stacking of thylakoids. As reported by S. Izawa and N. Good, when chloroplasts are suspended in H_2O the thylakoids become unstacked, separated and bloated. This derangement is reversible; complete restacking is effected by adding 3-5 mM Ca^{2+} or Mg^{2+}, or 50-100 mM Na^+. Photochemical activity remains normal throughout these gross morphological changes. These effects can be ascribed to electrostatic interactions of cations with negative charges on the thylakoid surfaces. At pH above 4.3 the surfaces carry a net negative charge, due mainly to carboxyl ($-COO^-$) groups of proteins and perhaps also to the hydrophilic head groups of phospholipids. Unless these charges are screened by positive ions, the membranes separate because of electrostatic repulsion. Cations neutralize the negative charge and allow the membranes to stick together, probably with the help of hydrophobic interactions (adhesion of water-repellent parts of the surfaces). If the free $-COO^-$ groups are replaced by glycine-methyl esters, permanent stacking is the result, even in H_2O. One can imagine that divalent cations are especially effective because they can link the negative sites on adjacent membranes, $(-)(++)(-)$, providing electrostatic bonds that supplement the hydrophobic attraction. This rationale has been pursued analytically by J. Barber and G. F. W. Searle, who demonstrated good quantitative agreement between theory and observation of the effects of mono- and divalent cations in screening surface charges and in promoting stacking.

If we ask which membrane components (if any) are specifically involved in stacking, we find that the evidence is suggestive but somewhat contradictory. In general it appears that stacking is correlated with the presence of Photosystem 2 and LHa/b. In normal chloroplasts that have stacked thylakoids, Photosystem 2 activity is confined largely to the stacked regions. In the thylakoid fragments, antibodies specific for LHa/b interfere with the ability of cations to promote stacking. In certain mutants of tobacco the absence of Photosystem 2 activity or of LHa/b is correlated with the absence or paucity of stacks. But in some mutants of pea and barley, and in bean leaves that have developed in intermittent (2 min on, 2 min off) light, there is no Chl b, no LHa/b, and no (or few) of the large 16 nm particles typical of stacked regions, and yet some stacking can be seen. And in other mutants there are stacks although Photosystem 2 activity (or, alternatively, Photosystem 1 activity) is missing. We are far from understanding this problem. It may be that "absence" of Photosystem 2 or LHa/b is an oversimplification. Minor polypeptides of these components, necessary for stacking, may be present when the components appear to be absent.

Significant changes of energy transfer and quantum distribution require only that pigments move a nanometer or less, toward or away from each other. The gross movements seen in stacking, through many hundreds of nanometers, may conceal subtler movements on which the control of quantum distribution depends. Indeed there are light-induced morphological changes less spectacular than stacking and unstacking, but far coarser than a nanometer's displacement. When chloroplasts are illuminated, the stacks of thylakoids become appressed more closely, shrinking about 25% and then relaxing in the dark. This effect, easily monitored by an attendant change in the scattering of light by the chloroplasts, depends on the translocation of ions, perhaps Mg^{2+}. Agents that render the thylakoids freely permeable to H^+ eliminate the light-induced shrinking; one can argue that Mg^{2+} efflux is abolished by such agents. With H^+ free to diffuse out of the thylakoid as rapidly as it is pumped in, there is nothing to cause Mg^{2+} to move across the membrane.

As for submicroscopic movements on the scale of a nanometer, we have only a few fragments of evidence suggesting changes in the orientations of Chl molecules in the membrane (see the next chapter), and of course the observation that patterns of energy transfer do change.

By 1970, scientists were confronted with a welter of information about fluorescence in chloroplasts. Observations to which we shall return in Parts III and IV included effects of light, pH, temperature, ions, electron carriers, agents that alter the permeability of membranes, and various inhibitors, taken singly and in sequences and combinations. Interpretations of these phenomena centered in the states of the reaction centers, in the electrochemical gradients (across the membranes) that mediate ATP formation, and in the patterns of energy transfer among pigment–protein components. There was an urgent need for unifying generalizations that could bring this accumulation of facts and interpretations into focus. To this end a number of investigators, especially W. L. Butler and his colleagues, undertook to isolate and interpret those aspects that involve the transfer of energy and its utilization by the reaction centers.

Butler's "tripartite" analysis was based on the presumed existence of the three components Core 1, Core 2, and LHa/b, and on the fact that leaves and chloroplasts show three distinctive fluorescence bands when brought to liquid nitrogen temperature, 77 K. These bands have maxima near 680, 695, and 735 nm. Isolated LHa/b, Fraction 2, and Fraction 1 exhibit fluorescence maxima near 680, 685, and 735 nm, respectively, at low temperature. Thus the fluorescence bands in the intact leaf can be associated with the three fractions, except that the 695 nm band in the leaf corresponds to the 685 nm band in purified Fraction 2.[6] Butler proceeded on the assumption, apparently well justified, that the three fluorescence bands of chloroplasts at low temperature are sufficiently well separated that the intensities of fluorescence measured at 680, 695, and 730 nm give the relative numbers of excitation quanta in each component. He formulated equations that expressed three essential aspects of the problem: the fraction of quanta absorbed initially by each component, the rates or probabilities of energy transfer from one component to another, and the rates of loss (by fluorescence, radiationless dissipation, and photochemical utilization) of excitation quanta in each component. The intensity of fluorescence from each component equals the concentration of quanta in that component multiplied by the rate constant for fluorescence. The possibility of multiple transfers between components was taken into account.

This kind of analysis could be applied to materials treated in various ways before freezing – with or without divalent cations, with illumination to produce State I or State II, with chemical oxidation or reduction to alter the states of reaction centers – and in mutants deficient in a component or a photochemical activity. After successive interactions between experiment and theory, Butler and his collaborators could draw several conclusions. Some of these conclusions, which are listed below, were reached through measurements of photochemical electron transfer and light-induced absorbance changes (to be considered in Part III) as well as from measurements of fluorescence.

1.   Core 2 is coupled closely to LHa/b, with rapid energy transfer between these components. Quanta are also transferred from LHa/b and Core 2 to Core 1; but once a quantum has reached Core 1, it is unlikely to return to the other components. Divalent cations (or "State I") increase the coupling between LHa/b and Core 2 and diminish the flow of quanta to

Core 1, thus providing one basis for the regulation of quantum distribution to the two photosystems.

2.   The rate of trapping of quanta by the reaction center of Photosystem 1 is almost independent of the state of the reaction center except under extreme conditions. The trapping rate appears to be unaffected by oxidation of the photochemical electron donor P700 (a special pair of Chl *a*). The two earliest electron acceptors in Photosystem 1 pass their electrons to secondary acceptors so quickly that the early ones do not accumulate in their reduced forms except in a strongly reducing environment with strong illumination (see Section 8.2). Reduction of the peripheral acceptors does not affect the rate of trapping by the Photosystem 1 reaction center. Thus one does not ordinarily see any variable component of fluorescence that can be correlated with photochemical activity in Photosystem 1. A minor exception is found in red and blue-green algae, where the rate of trapping is slightly lower when P700 is oxidized.

In Photosystem 2 the rate of trapping is also independent of whether the photochemical electron donor (another special pair of Chl *a*) is reduced or oxidized, as long as the electron acceptor (plastoquinone) is not reduced. But during illumination, the acceptor is driven to its reduced form while the donor is kept reduced by a rapid flow of electrons from water. In this condition the rate of trapping by the reaction center is diminished. With trapping by the reaction center less effective, the density of quanta builds up in Core 2 and consequently in LH*a/b* and Core 1. This is manifested mainly by increased fluorescence (Fig. 6.4) in the bands identified with Core 2 and LH*a/b*, but the increased spillover to Core 1 can be seen also, by a parallel but smaller increase in the 730 nm fluorescence. This variable component, $f_v(730)$, is a direct sign of a second regulatory device: When Photosystem 2 is overdriven, its reaction centers become "closed" as traps, and more quanta are diverted to Photosystem 1.

3.   Analysis of the constant and variable components $f(730)$ and $f_v(730)$ provides a basis for two kinds of action spectra (i.e., fluorescence excitation spectra). The excitation spectrum for $f_v(730)$ pertains to changes in quantum utilization in Core 2 and the closely coupled LH*a/b*. This spectrum shows a principal maximum at 677 nm attributed to absorption by Chl *a* in Core 2, and lesser maxima at 650 and 670 nm ascribed to Chls *b* and *a* in LH*a/b*. The total $f(730)$, corrected for spillover originating in Core 2, has an excitation spectrum showing a principal maximum at 681 nm (Chl *a* in Core 1) and lesser maxima at 650 and 670 nm (LH*a/b*).

4.   This interpretation of the fluorescence and its variations is internally consistent and can account for approximately balanced operation of the two photosystems, with constant overall efficiency at wavelengths below 680 nm. When plants are illuminated after prolonged dark adaptation, there is an initial excess of quanta to Photosystem 2, but the two regulatory devices soon correct the imbalance. One can speculate that the first regulatory device, the

change of coupling between the components, is based on light-induced Mg^{2+} efflux from the thylakoids. The second and more obvious device is based on the fact that when Photosystem 2 receives an excess of quanta, the trapping by its reaction center is attenuated, and quanta are shunted to Photosystem 1.

Initially Butler based his analysis of fluorescence on a model of the kind shown in Fig. 6.1*a*. This model was revised to the form of Fig. 6.1*b*, mainly because in plants that lack LH*a/b* (such as bean seedlings developing in intermittent light) the fluorescence at 730 nm has a variable component. This $f_v(730)$, a reflection of changes in Photosystem 2, showed that excitation energy can be transferred from Core 2 to Core 1 without going through LH*a/b*.

A natural criticism of Butler's development is that it is based on measurements made under nonphysiological conditions (liquid nitrogen temperature). For example, Butler acknowledges that the 730 nm fluorescence comes from a spectral form of Chl *a*, C705 (absorption maximum at 705 nm), that only becomes fluorescent as the temperature is lowered. Even so, this fluorescence is a valid indicator of quantum density in Core 1 at low temperature. The pattern of energy transfer among the three components at low temperature may differ from that at room temperature, but the more limited range of observations that can be made at room temperature (with only one distinct fluorescence band) is fully consistent with the general picture that has emerged from the measurements at low temperature.

Balanced operation of the two photosystems is not necessarily optimal for the growth of a plant. When the two photosystems act in concert, as in Fig. 3.4, $NADP^+$ is reduced by electrons from water, and some ATP is formed (this information will be developed in Chapter 9). A cyclic pattern of electron flow driven solely by Photosystem 1 can lead to further ATP formation without $NADP^+$ reduction. Both products, NADPH and ATP, are required for the assimilation of CO_2 by the reductive pentose cycle, as we shall see in Chapter 11. If the plant requires additional ATP, beyond what is made in the noncyclic operation of Photosystems 1 and 2 in series, it might be advantageous to supplement the noncyclic mode with a cyclic operation of Photosystem 1 alone. These considerations are speculative; it has not been established that the cyclic pattern demonstrable in suspensions of chloroplasts plays a significant role in intact leaves. It is well to keep this added dimension of flexibility in mind, however, when visualizing the network of relationships between H^+ pumping, ATP formation, Mg^{2+} efflux, and quantum distribution.

6.4 Antenna components and energy transfer in diverse types of algae and bacteria

Algae with chlorophyll *c* (marine dinoflagellates and diatoms)

The photosynthetic marine dinoflagellates have the morphology of protozoa encased in exoskeletons composed mainly of cellulose, with flagella

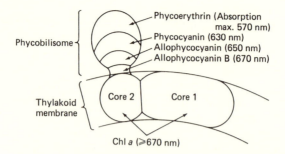

Fig. 6.5. A model for the organization of pigment–protein components in red and blue-green algae. The phycobilisomes in these algae appear to replace the function of LH$a/b$ (see Fig. 6.1) in other algae and plants.

extending through holes in their armor. Diatoms are tiny ellipsoids encapsulated in a silicious armor. These algae comprise a major component of the microscopic life of the coastal parts of the seas. They have the usual photochemistry of algae, Photosystems 1 and 2. Their major pigments are Chl a, Chl c (a type of Chl with absorption maxima near 450 and 630 nm), and the carotenoid peridinin. On the basis of isolated pigment–protein components it has been proposed that in these algae the organization of antenna components is like that of plants as shown in Fig. 6.1, but with a Chl a–Chl c-protein complex in place of LH$a/b$, and perhaps with a shell of Chl a–peridinin-protein around the "LH$a/c$."

Bacteria with chlorophyll a (prokaryotic algae)

The green prokaryotic algae of the genus *Prochloron* appear to have the usual Photosystems 1 and 2 of other algae and plants, and the familiar organization of antenna components based on Core 1, Core 2, and LH$a/b$. These unicellular algae should be classified as bacteria on the basis of their intracellular morphology, with no organized nucleus.

The prokaryotic blue-green algae or cyanobacteria share with red algae the distinctive "antenna" organelles called phycobilisomes (recall Section 2.6 and the introduction to Part II). These bodies are attached to the outer surfaces of the thylakoids, but are readily removed from cells by washing with aqueous buffer solutions. They serve a function comparable to that of LH$a/b$ in other algae; red and blue-green algae have no Chl b. A model consistent with known paths of energy transfer, patterned after the arrangement of Fig. 6.1b, is shown in Fig. 6.5. This arrangement provides for a "downhill" cascade of excitation energy through the phycobilins to Chl a. The phycobilins are coupled more closely to Core 2 than to Core 1, as with LH$a/b$ in relation to these components in other algae and plants, but there is copious spillover to

Core 1. Most of the Chl a should be associated with Core 1, most of the energy absorbed by phycobilins is transferred directly to Core 2, and about half or more of the energy received by Core 2 spills over to Core 1. The partitioning of quanta between Photosystems 1 and 2 is subject to the same types of regulatory devices that operate in tissues that have LH$a/b$. In addition there is chromatic adaptation: When such algae are grown in green light, absorbed mainly by phycoerythrin, they synthesize relatively more phycoerythrin; growth in orange light leads to the synthesis of relatively more phycocyanin.

Bacteria with bacteriochlorophylls a and c ("green" photosynthetic bacteria)

The bacteria that contain Bchl are divided broadly into green sulfur bacteria, with Bchl c as a major antenna pigment and Bchl a as a minor one, and "purple" bacteria, with Bchl a or Bchl b alone. The purple bacteria are divided further into sulfur bacteria (Thiorhodaceae) and nonsulfur bacteria (Athiorhodaceae) on the basis of secondary metabolic capabilities, primarily the ability to use sulfur as an oxidizable substrate. The use of colors in describing these categories of bacteria is unfortunate. Some "purple" bacteria are green because they contain yellow carotenoids and blue Bchl; most are red or brown or peach-colored; I have never seen a culture that looks purple.[7] Furthermore the alga *Prochloron* can be described properly as a green photosynthetic bacterium. We shall reserve the common names "green sulfur bacteria" for those that contain Bchl c and "purple bacteria" for those that contain only Bchl a or Bchl b.

The best-known genus of green sulfur bacteria is *Chlorobium.* The spurious species *"Chloropseudomonas ethylica"* has been a source of great taxonomic confusion. For many years it was not appreciated that cultures designated by this name were actually mixed (symbiotic) cultures of a green sulfur bacterium and a nonphotosynthetic sulfate reducing bacterium.[8] To make matters worse, "strain N2" of *"Chloropseudomonas ethylica"* was found to contain *Chlorobium limicola*, whereas "strain 2K" contains another green sulfur bacterium, *Prosthecochloris aestuarii.*

The green sulfur bacteria differ profoundly, in their intracellular structure and in the details of their photochemistry, from the purple bacteria. The photochemistry generates a strong reductant, of midpoint potential about -0.55 V, in contrast to the reductant of potential about -0.15 to -0.2 V made by purple bacteria. In this respect the reaction center of green sulfur bacteria is reminiscent of the Photosystem 1 reaction center in plants.

Green sulfur bacteria have a major antenna component of Bchl c, with two chemically distinct forms giving an absorption maximum at either 725 or 750 nm (only one form in any one species of bacterium). There is a subantenna of Bchl a with an absorption maximum near 810 nm, and the reaction center contains Bchl a (long wave absorption maxima at 830 and

842 nm) and bacteriopheophytin *a*. The molecular ratio of Bchl *c*:Bchl *a*: reaction centers is about 1000:80:1. The Bchl *c* is packaged in prolate ellipsoidal vesicles about 100 nm long, each containing about 10,000 molecules of Bchl *c*. These vesicles lie in profusion inside the cell against its periphery. Each vesicle is bounded by a "single" proteinaceous membrane, in contrast to the protein–lipid–protein bilayered structure ("unit membrane") found in cytoplasmic and intracytoplasmic membranes. Sandwiched between the vesicles and the cytoplasmic membrane, in a space about 8 nm thick, is a subantenna of Bchl *a*-protein. The reaction centers, and perhaps also some of the antenna Bchl *a*, are probably built into the cytoplasmic membrane. These assertions must be tempered by the admission that we know far less about structural details in green sulfur bacteria than in purple bacteria. In particular, it has not been proved, although it is very likely, that the reaction centers are integral components of the cytoplasmic membrane.

Taken at face value, the arrangement of photosynthetic components in green sulfur bacteria is reminiscent of that in blue-green algae. Major antenna components (the Bchl *c* vesicles and Bchl *a* subantenna) are bound loosely to a membrane containing the photochemical apparatus. In blue-green algae the phycobilisomes are bound loosely to the thylakoid membrane. In both cases these accessory light-harvesting pigments probably represent evolutionary maneuvers in which new antenna pigment has been superimposed upon a preexisting membrane containing antenna chlorophyll and reaction centers. Phycobilisomes are removed from thylakoid membranes by washing with aqueous buffer solutions. The Bchl *a*-protein subantenna of green sulfur bacteria is brought into aqueous solution, without the help of detergents, by breaking the bacteria and centrifuging away the vesicles and other particulate components.

The hydrophilic nature of the Bchl *a*-protein has allowed its crystallization (by J. M. Olson and collaborators), and R. E. Fenna and B. W. Matthews have determined its structure to a nominal resolution of 0.28 nm by X-ray diffraction. In crystals this Bchl *a*-protein complex exists as a trimer with seven molecules of Bchl *a* in each of the three subunits. The subunit has been described as a "string bag" of protein, weighing 39 kdalton, in which the seven Bchls are held in position by hydrophobic interactions. The subunit is about 2–4 nm in linear dimension, and the average center-to-center distance between neighboring Bchls is about 1.2 nm. The tetrapyrrole planes of the seven Bchls make angles of 10° to 40° with an "average plane." The amino acid composition is known, and the amino acid sequence will probably be available soon. Detailed absorption and circular dichroism (see Section 7.2) spectra have been measured; they provide a good test for the validity of theoretical calculations of such parameters. The calculated spectra, based on theories of interaction between the dipole oscillations that accompany transitions between ground and excited states of the Bchls (see Section 7.1), show fair but not excellent agreement with the observed spectra. The lack of perfect correspondence probably testifies to the complexity of the problem, necessitating simplifying assumptions, and not to intrinsic inadequacies of the theory. In summary, this Bchl *a*–protein, by virtue of its hydrophilic character, has become by far the most thoroughly characterized of all chlorophyll-protein complexes found in nature.

Photosynthesis without chlorophyll; the purple membranes of *Halobacterium halobium*

This remarkable bacterium lives in brackish marine backwaters, growing optimally if the water contains 25% NaCl, and smelling of old tennis shoes. It had been studied for decades before W. Stoeckenius discovered in 1967 that when grown in the light at a suboptimal concentration of oxygen, it develops purple patches in its cytoplasmic membrane. These oval patches, about 1 μm long, can occupy more than half the area of the membrane. The purple patches can be isolated by differential centrifugation of broken cells. Freeze-fracture, electron microscopy, and X-ray diffraction studies show that the patches have a crystalline lattice structure in which the unit cell is a trimer of pigmented protein molecules. The space around the protein is intercalated with phospholipids.

The individual pigment–protein molecules in the purple patches, called bacteriorhodopsin, are closely similar to the visual pigment rhodopsin found in the vertebrate retina: The chromophore, retinal (the aldehyde of vitamin A), is bound to an opsin-like protein (weight 26 kdalton) by a protonated Schiff base link to the ϵ-amino group of lysine:

$$\begin{array}{c} \text{Retinal} \\ \text{R}-\text{CHO} \end{array} + \begin{array}{c} \text{Lysine} \\ \text{H}_2\text{N}-\text{R}' \end{array} \xrightarrow[+\text{H}^+]{-\text{H}_2\text{O}} \begin{array}{c} \text{H} \\ \text{R}-\text{CH}=\overset{}{\underset{\oplus}{\text{N}}}-\text{R}' \end{array}$$

After illumination, the retinal is found in a *trans* isomeric configuration with an absorption maximum at 570 nm (extinction coefficient 63 mM^{-1} cm^{-1}). In the dark this pigment drifts over the course of minutes to a *cis* configuration absorbing maximally at 560 nm. The light-adapted (570 nm) form engages in a rapid photochemical cycle reminiscent of the cycle of rhodopsin in the vertebrate eye. The cycle involves deprotonation and reprotonation of the Schiff base link:

$$\begin{array}{c} \text{H} \\ \text{R}-\text{CH}=\overset{}{\underset{\oplus}{\text{N}}}-\text{R}' \end{array} \underset{+\text{H}^+}{\overset{-\text{H}^+}{\rightleftharpoons}} \text{R}-\text{CH}=\text{N}-\text{R}'$$

The deprotonated form has its absorption maximum at 412 nm.

Bacteriorhodopsin spans the cytoplasmic membrane, and during the photochemical cycle H$^+$ is ejected on the outside of the membrane and regained from the inside. Thus the purple patches mediate light-induced pumping of

H⁺ out of the cell (as in "ordinary" photosynthetic bacteria; see the introduc- tion to Part II), and this in turn mediates the formation of ATP which can be used to drive energy-requiring metabolism.

Quantitatively this "photosynthesis without chlorophyll" is a pale substi- tute for the photosynthesis based on chlorophyll. *H. halobium* cannot grow at the exclusive expense of light energy, but light can help it through difficult times when oxygen is scarce and respiration is deficient. Bacteriorhodopsin also mediates a phototactic response which enables the cells to avoid dark regions.

SUGGESTED READINGS

Anderson, J. M. (1975). The molecular organization of chloroplast thyla- koids. *Biochim. Biophys. Acta 416*, 191–235.

Thornber, J. P., Alberte, R. S., Hunter, F. A., Shiozawa, J. A. and Kan, K.-S. (1976). The organization of chlorophyll in the plant photosynthetic unit. *Brookhaven Symp. Biol. 28*, 132–48.

Bengis, C. and Nelson, N. (1977). Subunit structure of chloroplast Photosys- tem 1 reaction center. *J. Biol. Chem. 252*, 4564–9.

Delepelaire, P. and Chua, N.-H. (1979). Lithium dodecyl sulfate/polyacryl- amide gel electrophoresis of thylakoid membranes at 4°C: Characteriza- tion of two additional chlorophyll *a*–protein complexes. *Proc. Natl. Acad. Sci. U.S. 76*, 111–15.

Staehelin, L. A., Armond, P. A., and Miller, K. R. (1976). Chloroplast mem- brane organization at the supramolecular level and its functional implica- tions. *Brookhaven Symp. Biol. 28*, 278–315.

Arntzen, C. J., Armond, P. A., Briantais, J.-M., Burke, J. J., and Novitzky, W. P. (1976). Dynamic interactions among structural components of the chloroplast membrane. *Brookhaven Symp. Biol. 28*, 316–37.

Bonaventura, C., and Myers, J. (1969). Fluorescence and oxygen evolution from *Chlorella pyrenoidosa*. *Biochim. Biophys. Acta 189*, 366–83.

Murata, N. (1970). Control of excitation transfer in photosynthesis. IV. Kinetics of chlorophyll *a* fluorescence in *Porphyra yezoensis*. *Biochim. Biophys. Acta 205*, 379–89.

Izawa, S., and Good, N. E. (1966). Effect of salts and electron transport on the conformation of isolated chloroplasts. I. Light-induced scattering and volume changes. II. Electron microscopy. *Plant Physiol. 41*, 533–543, 544–552.

Barber, J., and Searle, G. F. W. (1978). Cation induced increase in chloro- phyll fluorescence yield and the effect of electrical charge. *FEBS Lett. 92*, 5–8.

Butler, W. L. (1978). Energy distribution in the photochemical apparatus of photosynthesis. *Annu. Rev. Plant Physiol. 29*, 345–78.

Prézelin, B. B., and Alberte, R. S. (1978). Photosynthetic characteristics and organization of chlorophyll in marine dinoflagellates. *Proc. Natl. Acad. Sci. U.S. 75*, 1801–4.

Withers, N. W., Alberte, R. S., Lewin, R. A., Thornber, J. P., Britton, G., and Goodwin, T. W. (1978). Photosynthetic unit size, carotenoids, and chlorophyll–protein composition of *Prochloron* sp., a prokaryotic green alga. *Proc. Natl. Acad. Sci. U.S. 75*, 2301–5.

Trüper, H. G., and Pfennig, N. (1978). Taxonomy of the Rhodospirillales. In *The Photosynthetic Bacteria*, R. K. Clayton and W. R. Sistrom, eds., Chapter 2, pp. 19–27. Plenum, New York.

van Niel, C. B. (1944). The culture, general physiology, morphology and classification of the non-sulfur purple and brown bacteria. *Bacteriol. Revs. 8*, 1–118.

Larsen, H. (1953). On the microbiology and biochemistry of the photosynthetic green sulfur bacteria. Doctoral thesis, The Norwegian Institute of Technology, Trondheim.

Olson, J. M. (1978). Bacteriochlorophyll *a*–proteins from green bacteria. In *The Photosynthetic Bacteria*, R. K. Clayton and W. R. Sistrom, eds., Chapter 8, pp. 161–78. Plenum, New York.

Oesterhelt, D., and Stoeckenius, W. (1973). Functions of a new photoreceptor membrane. *Proc. Natl. Acad. Sci. U.S. 70*, 2853–7.

Stoeckenius, W. (1979). A model for the function of bacteriorhodopsin. In *Membrane Transduction Mechanisms*, R. A. Cone and J. Dowling, eds., Society of General Physiologists Series, Vol. 23, pp. 39–47. Raven, New York.

7 Measurements with polarized light: interactions of molecules in excited states; orientations of pigments in photosynthetic tissues

7.1 Digression: theory and general methods involving polarized light

In an electromagnetic wave (Section 2.1 and Fig. 2.1) the electric and magnetic vectors are everywhere mutually perpendicular, proportional to each other in magnitude, and perpendicular to the direction of propagation. In classical theory the wave is generated by an oscillating dipole (emission) and can induce other dipole oscillations (absorption). When a molecule absorbs or emits light, the transient dipole oscillation has a direction and magnitude that can be computed, at least in principle, from the difference in the distribution of charge in ground and excited states. Because this redistribution of charge is spread out in the molecule, one is dealing with an "extended dipole" and not a single pair of + and − charges (point dipole). As an approximation, the extended dipole can be represented by an equivalent point dipole having a certain direction and strength. The strength of a point dipole is the product of charge and separation. A vector specifying the direction and strength is called the dipole moment $\overline{M}$; in the context of a radiative transition it is called the transition moment.

The various electronic transitions in a molecule, manifested as principal bands[1] in its absorption spectrum, correspond to natural frequencies of dipole oscillation, and each has its characteristic direction in the framework of the molecule. Absorption can be regarded as a resonance process in which the oscillating electric vector of the light wave induces the appropriate oscillation. Thus it is the component of the electric vector along the line of the dipole oscillation that determines the probability of absorption. A proper analysis shows that the squares of these parameters should be used in computing the probability. This requirement is related to the fact that the flux of energy in an electromagnetic wave (or the intensity of a beam of light) is proportional to the square of the electric vector.[2] Then if the transition moment $\overline{M}$ makes an angle θ with the electric vector $\overline{E}$, the probability of absorption is proportional to $E^2 M^2 \cos^2 \theta$. In this microscopic (single molecule) picture E^2 represents local light intensity and $M^2 \cos^2 \theta$ is related to the extinction coefficient ϵ for light with $\overline{E}$ making the angle θ with $\overline{M}$. We shall apply these considerations to oriented macroscopic systems presently.

In the case of emission, the electric vector of the emitted light has the direction of the corresponding dipole oscillation (transition moment, $\overline{M}$) in the molecule. Most sources of light consist of collections of emitting molecules oriented randomly. If the light is collimated into a parallel beam, the electric vector is perpendicular to the axis of propagation, oriented randomly around it. Various devices can reject all but a single orientation of $\overline{E}$, giving plane-polarized light. Some types of polarizers use crystals in which the light is refracted differently for different orientations of $\overline{E}$ relative to the axes of symmetry of the crystal (birefringent crystal). Other types are sheets of plastic or glass that contain light-absorbing molecules, oriented so that one polarization is absorbed and another (at right angles) is transmitted ("Polaroid"). Still others use the principle that the reflection of light from a surface (e.g., glass) depends on the polarization. In fact, many optical instruments cause the light to become polarized, and the experimenter should be aware of this. In a prism monochromator the light is usually incident on the face of the prism at an angle close to the one (Brewster's angle) that gives maximum polarization by selective reflection.

Consider a beam of plane polarized light viewed along the axis of propagation, as in the upper part of Fig. 7.1. Coordinate axes X and Y are chosen so that the plane of polarization lies midway between them. The components E_x and E_y are in phase, and $\overline{E}$ varies sinusoidally in magnitude but stays in its plane. Now interpose a birefringent crystal such that the indices of refraction are different for the X and Y components of $\overline{E}$. If the thickness of the crystal is chosen so as to retard E_x by a quarter wave (90°) relative to E_y, or vice versa, the result is circularly polarized light in which $\overline{E}$ describes a circle as seen along the axis of propagation. This can be appreciated by studying Fig. 7.1 and, if necessary, making sketches of E_x and E_y vs. time. If the retardation is 180°, the light remains plane-polarized but is reoriented perpendicular to the original direction. Retardation by some other amount gives elliptically polarized light. Thus one can build an instrument giving right or left circular polarization, or one that presents these types alternately, or one that gives plane-polarized light alternating in its orientation. There are electro-optical devices in which the retardation can be controlled and switched by applying alternating voltage to the device.

For right circularly polarized light ($\overline{E}$ rotating clockwise as seen along the axis of propagation from behind) the electric vector sweeps out a helix along the path of propagation, with the sense of a right-handed screw. For left circular polarization $\overline{E}$ describes a left-handed helix. If a molecule has helical structure in those parts that are involved in electron redistribution during a transition, it will show preferential absorption for right or left circularly polarized light depending on the sense (right or left) of the helix. This will happen even in a solution of randomly oriented molecules because a right-handed helix looks right-handed from either end. As a general criterion, a

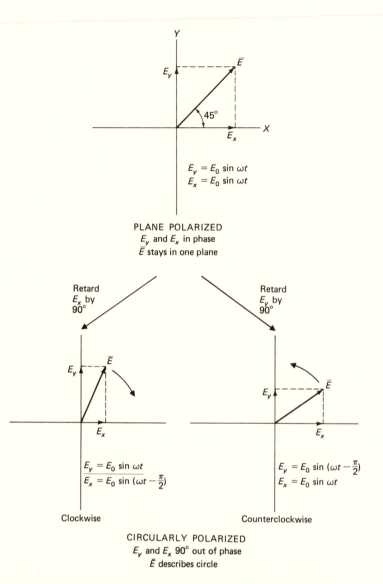

PLANE POLARIZED
E_y and E_x in phase
$\bar{E}$ stays in one plane

$E_y = E_0 \sin \omega t$
$E_x = E_0 \sin \omega t$

Retard E_x by 90°

Retard E_y by 90°

$E_y = E_0 \sin \omega t$
$E_x = E_0 \sin (\omega t - \frac{\pi}{2})$

$E_y = E_0 \sin (\omega t - \frac{\pi}{2})$
$E_x = E_0 \sin \omega t$

Clockwise

Counterclockwise

CIRCULARLY POLARIZED
E_y and E_x 90° out of phase
$\bar{E}$ describes circle

Fig. 7.1. A beam of light is viewed along the axis of propagation (perpendicular to the drawing). In the upper sketch the light is polarized in a plane 45° from the X and Y axes. At any given point along the axis of propagation the electric vector $\bar{E}$ varies sinusoidally in magnitude but stays in this plane. Quarter-wave (90°) retardation of the x or the y component gives circularly polarized light as shown. With half-wave retardation (180°) of E_x or E_y the light would remain plane-polarized, but in a plane perpendicular to the original one.

molecule will show aspects of helical structure (called chirality) if it cannot be superimposed on its mirror image. An organic molecule has chirality if a carbon atom in the molecule is bonded to four different residues:

$$R_2 - \underset{\underset{R_3}{|}}{\overset{\overset{R_1}{|}}{C}} - R_4$$

A circular dichroism (CD) spectrometer, which measures the difference between absorption of right and left circularly polarized light, is a sensitive instrument for detecting the degree and the sense of chirality, and thus for probing molecular structure. A simpler instrument, the polarimeter, has long been used to measure another aspect of chirality: the ability of helical structures to rotate the plane of plane-polarized light. This is a convenient way to determine the concentrations of strongly chiral substances, such as sugars, in solution.

A molecule can acquire new and stronger chirality when it becomes associated as a dimer, such that the excited state of the dimer is the joint property of both monomer subunits. The quantum of excitation is delocalized over both monomers; one speaks of strong coupling (strong electric dipole coupling, strong exciton coupling, etc.; see the last paragraph of Section 2.1). A quantum that is thus delocalized over two or more molecules is called a molecular exciton. We shall see that a single excited state of a monomer gives two states in the dimer. In general the energies of these two states are not equal; the dimer shows two absorption bands for each one in the monomer. Often the difference in energy is so small that the bands are not well resolved in absorption, but the resolution is clear in a CD spectrum. This is because the chirality induced by the dimeric association is of opposite sense for the two transitions. The CD can also distinguish between a dimer and two noninteracting monomers whose excited state energies differ slightly because of different environments. These points are illustrated in Fig. 7.2. In summary, CD can be used to probe the strength of intermolecular electric dipole coupling as well as intramolecular chirality.

Strong electric dipole interactions between pigment molecules can be described in a simple graphic way, based on the theory of molecular excitons as developed by A. S. Davidov and extended by I. Tinoco and M. Kasha. In this simple description the dipoles are treated as point dipoles. Consider a monomer with a transition moment $\overline{M}$, and imagine two such monomers situated in space as shown in Fig. 7.3d, where the transition moments $\overline{M}_1$ and $\overline{M}_2$ of the separate monomers are identical except for their location and orientation. These monomers form a dimer, exposed to light of wavelength much greater than the dimensions of the dimer.[3] At any instant, then, the entire dimer is in the same phase of the wave. Figure 7.3d shows two modes

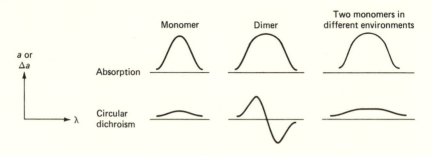

Fig. 7.2. The use of CD spectra in the discrimination of monomers and dimers; comparison with absorption spectra. (See text.)

of oscillation of the system, for two orthogonal orientations of the dimer relative to the electric vector $\bar{E}$. The mode shown on the left has "head-to-tail" polarity for the dipole oscillations in the two monomers; the other mode (right) has "head-to-head" polarity. In this presentation $\bar{M}_1$ and $\bar{M}_2$ are vectors with the positive sign arbitrarily given one direction as shown on the left in Fig. 7.3d. The head of the vector (arrow) is the positively charged end at any instant. If the arrowhead is reversed, the sign is also reversed, as for $\bar{M}_2$ on the right in Fig. 7.3d. The graphic representation as shown is then consistent with the analytical representation, $\bar{M}_1 + \bar{M}_2$ (left) and $\bar{M}_1 - \bar{M}_2$ (right). The dimer has two orthogonal transition moments given by $(\bar{M}_1 \pm \bar{M}_2)/\sqrt{2}$; the factor $1/\sqrt{2}$ gives the resultant its correct magnitude.[4] The mode shown on the left in Fig. 7.3d is an attractive configuration, and the mode on the right is a repulsive one, as can be appreciated by imagining the forces that such dipoles exert on each other. The attractive and repulsive modes correspond to two distinct excited states Exc_1 and Exc_2, as shown in Fig. 7.3a (the excited state energy of the monomer lies between these two). The separation in energy of Exc_1 and Exc_2 measures the strength of the electric dipole interaction between the monomers. The disposition of the monomers shown in Fig. 7.3d was chosen to give equal magnitudes of $\bar{M}_1 \pm \bar{M}_2$ for the two modes in the dimer. The two absorption bands corresponding to Exc_1 and Exc_2 then have equal intensity, as shown in the absorption spectrum of

Fig. 7.3. (a) Energies of ground and excited states of a monomeric pigment, and of two such monomers interacting as a dimer. (b) Absorption spectra, and (c) circular dichroism spectra for the monomer and the dimer. (d) Monomer transition moments and their vector summation to give the transition moments of the dimer, as described in the text. (e,f) Examples of this vector summation for two additional geometries (relative positions and orientations of the monomers).

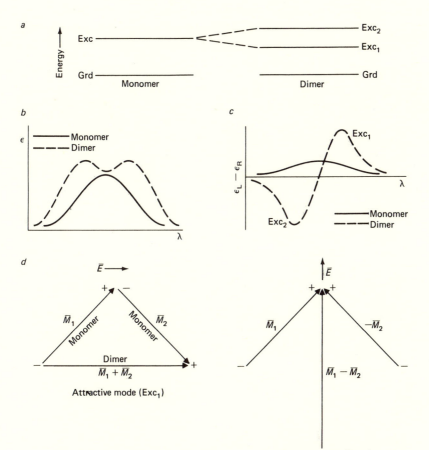

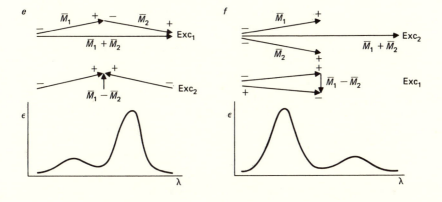

Fig. 7.3*b*. The CD spectrum (Fig. 7.3*c*) shows opposite (right or left) chirality for the two modes. With other relative orientations of the monomers the resultant $\overline{M}_1 \pm \overline{M}_2$ in the dimer has a different magnitude for each mode, but the positive and the negative CD bands are of equal amplitude. Figure 7.3*e* shows a case where $\overline{M}_1$ and $\overline{M}_2$ are nearly parallel and collinear. The transition moment of the dimer is then larger in the attractive mode than in the repulsive mode, and the absorption band of greater wavelength is the more intense. The reverse pattern is obtained when the monomer transition moments are nearly parallel and side by side (Fig. 7.3*f*). The foregoing simple description can be extended to more complicated situations involving the interactions of two or more molecules, with the monomeric transition moments not necessarily in the same plane.

The equations that underlie this pictorial description can be put succinctly in vector form as follows. In these equations the vectors $\overline{M}_1$ and $\overline{M}_2$ are the monomeric transition dipole moments (of equal magnitude; $M_1 = M_2 = M$), $\overline{R}_{12}$ is a vector joining the centers of the two dipoles, $D_\pm$ (i.e., D_+ and D_-) are the dipole strengths (absorption band intensities) of the dimer-split bands, $R_\pm$ (i.e., R_+ and R_-) are the rotational strengths or intensities of the positive and negative CD bands, and $\nu_\pm$ are the frequencies (at the absorption maxima) of the two bands. For the monomer, $D_0 = M^2$, $R_0 = 0$ (assuming no intrinsic chirality in the monomer), and the absorption maximum is at frequency ν_0.

$$D_\pm = M^2 \pm \overline{M}_1 \cdot \overline{M}_2$$
$$R_\pm = \mp \tfrac{1}{2}\pi\nu_0(\overline{R}_{12} \cdot \overline{M}_1 \times \overline{M}_2) \tag{7.1}$$

and

$$\nu_\pm = \nu_0 \pm \frac{1}{hcR_{12}{}^3}\left[\overline{M}_1 \cdot \overline{M}_2 - \frac{3}{R_{12}{}^2}(\overline{R}_{12} \cdot \overline{M}_1)(\overline{R}_{12} \cdot \overline{M}_2)\right]$$

In a single molecule the various electronic transitions, appearing as bands in an absorption spectrum, correspond to different symmetries of charge redistribution, each with its own characteristic dipole strength, orientation within the molecular framework, and frequency of oscillation. If a molecule contains an elliptical pattern of conjugated double bonds, with "π" electrons delocalized over these bonds, there will be two principal modes of oscillation with the dipole moment aligned with the long and short axes of the ellipse:

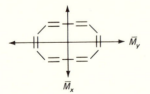

These modes correspond to two distinct $\pi\pi^*$ transitions. Such pairs of transitions can be seen in the spectra of monomeric chlorophylls, but additional bands arise from various complexities: nuclear vibrations, departures from perfect elliptical symmetry, participation of electrons that are not in the main set of conjugated double bonds, and multiple π and π^* states that correspond to different symmetries of the electron distribution in the main conjugation system. A rigorous accounting of the spectra of chlorophylls and other tetrapyrroles has not yet been made. We shall return briefly to the spectra of Chl a and Bchl a in the next section.

For a molecule with a given structure, axes can be defined in terms of that structure. In Chl, for example, with four N atoms in the tetrapyrrole part, the two N $\cdots$ N diagonals define X and Y axes, with the X axis pointing toward the corner where the hydrocarbon "tail" is attached (see Fig. 1.1). The orientation of a molecule in an external frame of reference can then be specified by giving the angular coordinates of the molecular axes relative to the external axes. If a collection of molecules of one type can be prepared so that every molecule has the same orientation in an external frame of reference, then the orientations of transition moments in the molecule can be probed with plane-polarized light.[5]

The absorption of light in a particular absorption band will be greatest when the electric vector of the light is aligned with the appropriate transition moment in each molecule. Usually one relies on theory to relate transition moments to molecular axes. Measurements of the absorption of plane-polarized light then give information about the angular coordinates of the molecular axes in the external frame of reference of the oriented sample. This technique is called linear dichroism spectrometry.

Several methods have been devised to obtain oriented samples of pigment molecules; a partial list follows.

1. The natural orientation in crystals. The chromophoric group may be all or part of the major component or may be introduced as a minor constituent. Some liquids also have aspects of crystalline order.

2. Adsorption onto films of materials such as cellophane.

3. Orientation of molecules as monolayers on liquid surfaces or at interfaces of immiscible liquids, and in monolayers or multilayers of lipids.

4. Hydrodynamic orientation (shear), as in a thin shell between two concentric cylinders with one rotating relative to the other.

5. Orientation in electric or magnetic fields, due to permanent or induced electric or magnetic dipoles in pigmented structures.

6. Orientation in films. Solutions of pigments mixed with polyvinyl alcohol or gelatin are dried to form films; the pigment may show planar orientation. If the film is humidified, stretched, and dried again, the pigment may become oriented axially (along the axis of stretching).

7. Orientation of biological membranes by drying membrane fragments onto plates, so that the fragments lie flat.

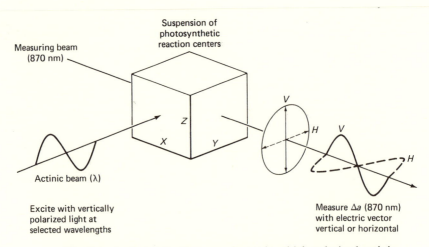

Fig. 7.4. A photoselection experiment in which polarized actinic light causes photochemistry and thus elicits a change of absorbance in a suspension of photosynthetic reaction centers. The orientation of the altered population is probed with a polarized measuring beam. (See further discussion in the text.)

8. Orientation by photoselection – a special category that will be discussed separately.

In recent years the last four of these methods have been applied extensively to components of photosynthetic tissues.

The method of photoselection has found desultory application in photosynthesis research for decades; its use has become more common recently. A randomly oriented collection of molecules is exposed to plane-polarized light so as to excite preferentially those molecules whose transition moments are aligned with the electric vector. Some consequence of this selective excitation is then tested for its polarization. An example is illustrated in Fig. 7.4. The sample cell contains bacterial reaction centers in aqueous suspension. Various transitions of Bchl and bacteriopheophytin are excited by vertically polarized actinic light, and the consequent absorbance change at 870 nm (a reversible bleaching due to oxidation of the Bchl special pair) is measured, with the measuring beam polarized first vertically and then horizontally. With 870 nm actinic light the change is greatest with vertically polarized measuring light; with 600 nm actinic light the change is greatest with horizontally polarized measuring light. A quantitative analysis of the data shows that the transition of Bchl excited by 600 nm light is perpendicular to that induced by 870 nm light.

The same technique can be used to measure the polarization of fluorescence emitted by the reaction centers when they are excited with polarized light. Fluorescence (downward transition) is polarized with the electric vector aligned with the transition moment, which has the same direction as the dipole moment of the corresponding upward transition. In a molecule with several distinct electronic transitions, with differently oriented transition moments, excitation into a shorter wave band usually yields emission from the longest wave transition (see Fig. 2.4). An exception may be found when a large molecule has two or more distinct chromophoric groups that are remote from each other and behave like independent "submolecules."

In general, any consequence of excitation can be explored by photoselection. For example, a collection of pigment molecules in a rigid medium can be subjected to irreversible photodestruction, using polarized light to bleach a selected fraction of the population. The surviving molecules are then oriented preferentially with their transition moments (for the particular transition induced by the actinic light) perpendicular to the electric vector of the actinic light. This can even be done with unpolarized actinic light because the surviving molecules will have their transition moments aligned preferentially with the axis of propagation of the actinic light; along this axis the electric vector is zero. Selective photodestruction thus yields an oriented sample that can be examined later for linear dichroism.

When using photoselection one must beware of several potential sources of error that diminish the selectivity:

1. In a turbid sample, light can be depolarized by scattering.
2. Energy transfer among randomly oriented molecules will randomize the excited population, even though the light was absorbed initially by a set with selected orientation.
3. A similar randomization will result from rotation of the molecules, during the time between the excitation and the expression or observation of the measured effect (e.g., fluorescence or photochemical alteration).
4. If the actinic light is so strong as to approach a saturating level, as for a photochemical process, most of the molecules in the sample will be affected, even those whose transition moments are far from being aligned with the electric vector of the actinic light. This washes out the photoselection.

The last two of these sources of depolarization are more severe when the measured effect is long-lived. With short-lived fluorescence, or with a photochemical change that is rapidly reversible, there is less time for rotation between excitation and observation. Also the intensity of actinic light needed for saturation is greater. If the photochemical change is irreversible, the molecules must be held in place until the sample has been probed.

Depolarization due to energy transfer can be diminished by dilution if the pigment is dispersed in solution. On the other hand, the degree of this source of depolarization can be used to estimate the extent ("number of jumps") of energy transfer. Depolarization due to light saturation can be eliminated by using actinic light that is weak enough (or, for an irreversible change, a small-enough integrated dose) to keep the response far short of maximal. Rotational depolarization is controlled by suspending the sample in a suitably viscous medium.

The half-time τ for rotational depolarization[6] is given by

$$\tau = \eta V/kT \tag{7.2}$$

where V is the volume (equivalent sphere) of the pigmented particle in cubic centimeters, η is the coefficient of viscosity of the suspending medium in $g\,cm^{-1}\,sec^{-1}$ or poise, k is Boltzmann's constant (1.4×10^{-23} joule deg^{-1}), and T is the temperature in degrees Kelvin. For a spherical particle of specific gravity $1.3\,g\,cm^{-3}$ weighing 8×10^4 daltons (an approximation to reaction centers from *Rhodopseudomonas sphaeroides*), $V = 10^{-19}\,cm^3$ (diameter 6 nm). If such particles are suspended in water at room temperature ($\eta = 10^{-2}$ poise), $\tau = 3 \times 10^{-8}$ sec. Rotational randomization should then be negligible during the $<10^{-9}$ sec lifetime of the fluorescence but should be severe for a photochemically altered state that is measured on the time scale of microseconds or longer. The viscosity of glycerol is 1000 times that of water, and cooling increases the viscosity enormously. Thus it is easy to extend the time scale over a wide useful range.

By observing the rate of rotational randomization in a photoselection experiment, one can measure the rotational mobility of a component in a biological membrane. Alternatively, one can estimate the size of a particle in suspension, or if the size is believed to be known, one can estimate its departure from a spherical shape.

As a final caution, one may sometimes have unrecognized prior orientation in a sample to be studied by photoselection, or unrecognized photoselection when measuring the linear dichroism of a light-induced change of absorbance in an oriented sample.

The analysis of linear dichroism and photoselection experiments is based on straightforward applications of analytic geometry and calculus, in which the factor $\overline{E} \cdot \overline{M}$ or $EM \cos \theta$ (see earlier in this chapter) is averaged appropriately over the entire set of molecules in a given situation. The resulting formulas give relationships between dichroic ratios, such as Δa(vertical)/Δa(horizontal) in Fig. 7.4, and angles made by transition moments with other transition moments or with coordinate axes. A few conclusions pertaining to commonly encountered situations are presented here; more general conclusions can be found in the suggested reading for this chapter.

For a sample oriented with axial symmetry, so that a transition moment $\overline{M}$ makes an angle θ with the axis of symmetry but is disposed randomly around that axis (Fig. 7.5, top sketch),

$$\epsilon_z = 3\epsilon_0 \cos^2 \theta$$

$$\epsilon_{xy} = (\tfrac{3}{2}) \epsilon_0 \sin^2 \theta \tag{7.3}$$

Axially symmetric orientation; linear transition moment M:

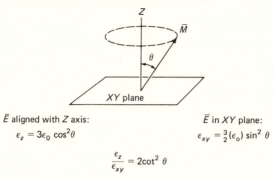

$\bar{E}$ aligned with Z axis:

$$\epsilon_z = 3\epsilon_0 \cos^2\theta$$

$\bar{E}$ in XY plane:

$$\epsilon_{xy} = \tfrac{3}{2}(\epsilon_0) \sin^2\theta$$

$$\frac{\epsilon_z}{\epsilon_{xy}} = 2\cot^2\theta$$

Axially symmetric orientation; circularly degenerate (planar) transition moment:

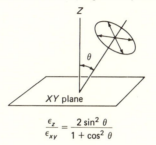

$$\frac{\epsilon_z}{\epsilon_{xy}} = \frac{2\sin^2\theta}{1 + \cos^2\theta}$$

If $\bar{E}$ makes angle α with Z axis:

$$\frac{\epsilon_\alpha}{\epsilon_{xy}} = \sin^2\alpha + \left(\frac{\epsilon_z}{\epsilon_{xy}}\right)\cos^2\alpha$$

(Extinction coefficient = ϵ_0 for random orientation)

Fig. 7.5. Formulas applicable to two specific cases of linear dichroism. (See discussion in the text.)

and

$$\epsilon_z/\epsilon_{xy} = 2\cot^2\theta$$

Here ϵ_0 is the extinction coefficient (at a wavelength that excites the transition whose moment is shown) for random orientation, ϵ_z is the extinction coefficient measured with $\bar{E}$ polarized along the axis of symmetry (the Z axis), and ϵ_{xy} is the extinction coefficient measured with $\bar{E}$ in the XY plane.

In some molecules there is a plane of symmetry such that transition moments in that plane can be decomposed into two orthogonal and equal moments of the same frequency (or energy). These degenerate moments give rise to a single absorption band; one speaks of a planar or circularly degenerate transition moment. For that case, with axially symmetric orientation (Fig. 7.5, center sketch) and with θ the angle between the Z axis and the normal to the plane of degeneracy,

$$\epsilon_z = (\tfrac{3}{2}) \epsilon_0 \sin^2 \theta$$

$$\epsilon_{xy} = (\tfrac{3}{4}) \epsilon_0 (1 + \cos^2 \theta) \tag{7.4}$$

and

$$\epsilon_z/\epsilon_{xy} = 2 \sin^2 \theta/(1 + \cos^2 \theta)$$

In applying these equations one measures absorbance or an induced change of absorbance (a or Δa) rather than ϵ or $\Delta\epsilon$. These parameters are proportional to one another, $a_z/a_{xy} = \epsilon_z/\epsilon_{xy}$ or $\Delta a_z/\Delta a_{xy} = \Delta\epsilon_z/\Delta\epsilon_{xy}$, provided that the optical path through the sample has the same length in each measurement. If not, the ratio of absorbances must be divided by the ratio of path lengths to give ratios of ϵ or $\Delta\epsilon$.

If axial orientation is achieved by drying membrane fragments flat onto a plate, it is technically unfeasible to have $\overline{E}$ aligned with the Z axis; for this the beam must pass through the sample edgewise. At best one can have the axis of propagation make a small angle with the XY plane, and this angle becomes greater inside the sample because of refraction.[7] One can measure ϵ_{xy} (axis of propagation along the Z axis; $\overline{E}$ in the XY plane) and ϵ_α (axis of propagation inside the sample making an angle α with the Z axis; $\overline{E}$ making an angle α with the XY plane); then (Fig. 7.5, bottom):

$$\epsilon_\alpha = \epsilon_z \cos^2 \alpha + \epsilon_{xy} \sin^2 \alpha$$

or $\tag{7.5}$

$$\epsilon_\alpha/\epsilon_{xy} = \sin^2 \alpha + (\epsilon_z/\epsilon_{xy}) \cos^2 \alpha$$

Formulas analogous to these are applicable when orientation has been achieved by photoselection, with the difference that perfect orientation is never achieved: For a randomly oriented sample in an XYZ coordinate system, light polarized along the Z axis does not excite exclusively the Z-oriented transition moments while leaving the others alone; transition moments inclined to the Z axis are excited in proportion to the cosine of the angle of inclination. If a single transition is excited with Z-polarized light and measured with Z- or XY-polarized light, as in Fig. 7.4, the ratio ϵ_z/ϵ_{xy} (or I_z/I_{xy} if fluorescence is measured) has a theoretical value of 3 if there are no sources of depolarization. If the transition moment involved in excitation is perpendicular to the one involved in measurement, the limiting (minimum) value of ϵ_z/ϵ_{xy} is 0.5. In linear dichroism, if one approaches perfect axial alignment of the transition moments with the Z axis, the ratio ϵ_z/ϵ_{xy} becomes indefinitely large, and if the transition moments are aligned perfectly in the XY plane, ϵ_z/ϵ_{xy} is zero (Equation 7.3).

Dichroism is often expressed not as a ratio such as ϵ_z/ϵ_{xy} but as a polarization factor P. In the present context P is defined as $P = (\epsilon_z - \epsilon_{xy})/(\epsilon_z + \epsilon_{xy})$. In general, if "$\parallel$" and "$\perp$" are used to denote measurement with polarization parallel to or perpendicular to that of excitation,

$$P = \frac{\epsilon_\parallel - \epsilon_\perp}{\epsilon_\parallel + \epsilon_\perp} \quad \text{or} \quad \frac{I_\parallel - I_\perp}{I_\parallel + I_\perp} \tag{7.6}$$

for measurements of absorption and emission, respectively. The limiting ratios 3 and $\frac{1}{2}$, predicted for $\epsilon_\parallel/\epsilon_\perp$ in photoselection when transition moments for excitation and measurement are mutually parallel or orthogonal, correspond to $P = \frac{1}{2}$ and $-\frac{1}{3}$, respectively. In a photoselection experiment that involves a circularly degenerate transition in either excitation or measurement, P cannot exceed $\frac{1}{7}$ and $\epsilon_\parallel/\epsilon_\perp$ can be no greater than $\frac{4}{3}$.

It is useful to note that in a measurement of linear dichroism one finds the angles that various transition moments make with one or more axes of symmetry, whereas with photoselection one computes the angles between pairs of transition moments. Further details about orientation can be learned by hybridizing the information from both techniques; an example of this will be given in the next section.

We have seen how the electric field of light can be used to probe orientations and interactions of electronic transitions in and among molecules. By another set of methods, using microwave spectroscopy, magnetic fields can be used to explore the orientations and interactions of the magnetic moments of unpaired electron spins in radicals and triplet-state molecular systems (recall Section 4.1). A few simple points will be listed here; for a detailed account of theory and applications the reader should consult books by Swartz et al. and Feher (suggested reading).

In conventional ESR spectroscopy, the sharing of unpaired electrons by more than one molecule is signaled by a decrease in the widths of the microwave absorption or emission bands. For an organic radical such as $Chl^{+\cdot}$ (spin = $\frac{1}{2}$), broadening of the ESR band is due primarily to interactions with magnetic nuclei (mainly protons but also ^{14}N and a trace of ^{13}C). Theory predicts, and observation supports the prediction, that if the unpaired electron is not confined to one molecule but is shared equally by n molecules, the band width is diminished by a factor $\sqrt{n}$. Comparison of band widths for oxidized Chl a or Bchl a in vivo with those for the monomeric cation radical in vitro has provided the primary evidence that the photochemical electron donors in reaction centers are "special pairs" of chlorophylls, except possibly in Photosystem 2.

Nuclear magnetic resonance (NMR) is analogous to ESR, but with radio frequencies (instead of microwaves) used to induce transitions between nuclear spin states. In electron–nuclear double resonance (ENDOR), the transitions between nuclear spin states are monitored by their effects on an ESR band. This gives well-resolved information about the magnetic nuclei (especially ^{1}H and ^{14}N) in the vicinity of an unpaired electron. The "special pair" hypothesis for reaction center chlorophylls has been confirmed by ENDOR.

ESR spectra of triplet states contain information ("zero field splitting parameters"; separations of distinctive bands) about the magnetic environment and interactions of the two unpaired electrons that give the triplet state its spin of $s = 1$. The information includes the approximate size and shape of the region occupied by these electrons, and thus can show whether the triplet state is delocalized over more than one molecule. The zero field splitting parameter D varies inversely with the third power of the average separation of the two unpaired electrons. When the triplet state is shared by two molecules, the size and shape of the domain of the unpaired electrons can suggest the relative positions and orientations of the individual molecules.

Finally, because the magnetic moments (or the spin axes) of the unpaired electrons are oriented relative to the molecular axes in addition to being orientated by the applied magnetic field, the methods of linear dichroism can be applied to oriented samples. Photoselection can also be used if light induces a magnetic species such as the oxidized special pair or the triplet state in reaction centers.

7.2   Spectroscopy of chlorophylls; measurements with polarized light as applied to photosynthetic tissues

In chlorophylls, X and Y molecular axes are chosen to lie in the tetrapyrrole plane along the N $\cdots$ N diagonals, with the X axis pointing toward the hydrocarbon tail of the molecule (upper right to lower left in Fig. 1.1). Absorption spectra of chlorophylls in solution (see Fig. 7.6) show two main pairs of bands, one pair in the blue-violet or near ultraviolet (the Soret bands) and one pair in the red or near infrared. Calculations based on theory indicate that these four bands reflect $\pi\pi^*$ transitions, polarized along either the X or the Y axis and having different symmetries (different combinations of distinct π and π^* states). For monomeric Chl a in ether, the maxima of the main bands are near 430 and 660 nm; for Bchl a in ether, they are near 360 and 770 nm. There are lesser bands with maxima near 410 and 575 nm in Chl a and near 390 and 575 nm in Bchl a. All of these bands are for the predominant (lowest-energy) modes of vibration in both ground and excited states. Other minor bands in these spectra are ascribed to transitions into higher vibrational modes.

The Soret bands have been designated B and the long wave bands Q. Using subscripts x and y to designate polarization, and (0, 0) or (0, 1) to show transitions from the lowest major vibrational mode in the ground state to the lowest (0) or the next one (1) in the excited state, we can catalogue the features in Fig. 7.6 by wavelength (see Table 7.1). Orthogonality of Q_x and Q_y has been verified for these chlorophylls, especially by measurements of polar-

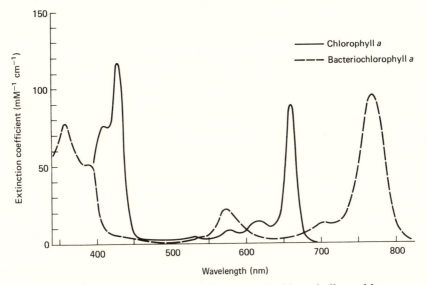

Fig. 7.6. Absorption spectra of monomeric chlorophyll *a* and bacteriochlorophyll *a* in ether.

Table 7.1. *Classification of principal transition in Chl a and Bchl a*

	Chl *a* (nm)	Bchl *a* (nm)
$B_y(0, 0)$	410	360
$B_x(0, 0)$	430	390
$Q_x(0, 0)$	575	575
$Q_y(0, 0)$	660	770
$Q_x(0, 1)$	530	535
$Q_y(0, 1)$	615	700

ized fluorescence in photoselection experiments. Information about the Soret (B_x, B_y) transitions is less complete than that about the Q_x and Q_y transitions for at least two reasons. First, one cannot detect fluorescence corresponding to the Soret bands, but only from the Q_y band. Second, the Q_x and Q_y bands and their vibrational substructures are well resolved in absorption spectra, whereas the B_x and B_y counterparts are not. Nevertheless the main features in the Soret region have been assigned with confidence to X or Y axes by photoselection, measuring the fluorescence. These conclusions hinge on the assumption, based primarily on theoretical predictions (molecular orbital calculations) of electron distribution, that the longest wave (fluorescent) transition is Y-polarized.

The extinction coefficients at the maxima of the main bands (Fig. 7.6) are about $100 \, \text{mM}^{-1} \, \text{cm}^{-1}$ for both Chl a and Bchl a. The Q_x band is more widely separated in energy from Q_y in Bchl a than in Chl a; this is consistent with the fact that the main system of conjugated double bonds is more elongated, on the average, in Bchl a than in Chl a.

The absorption spectra of chlorophylls in vitro, in various solvents and in different states of aggregation (dimer, etc.), show the perturbing effects of the environment of the individual Chl molecules. These perturbations have received extensive scrutiny in efforts to understand the organization (the nature and geometry of the materials surrounding each Chl molecule) in the living tissue. The most recent and thorough studies of Chl and Bchl in vitro have been made by J. J. Katz and his collaborators, and by K. Sauer in the area of circular dichroism.

The Q_y transition in Chl is perturbed more strongly by environmental influences than are the other major transitions. The principal environmental factors are electric dipole interactions. Pigment molecules can have permanent dipole moments that are different in their ground and excited states, and in the excited molecules, with their electron distributions more widespread, dipoles are induced more readily by nearby electric forces. For these reasons the differences in energy between ground and excited states depend markedly on the electric polarizability (dielectric constant) of the solvent, and on the charge distribution in the solvent (greater permanent dipole moment in polar solvents). The usual effect of these dipoles is to stabilize the excited state and lower its energy relative to the ground state, causing a shift of the absorption band to greater wavelengths.

When Chl molecules are brought close together in dimers and higher aggregates, their spectra are changed as a result of excitonic (electronic transition dipole) interactions.

The Mg atom has six coordination positions or sites of electrostatic bonding; in chlorophylls two of them are perpendicular to the tetrapyrrole plane. These are binding sites for polar components of the solvent, especially H_2O in "wet" solvents, and such polar groups can act as ligands in the formation of dimers and multimers of chlorophylls. In completely dry nonpolar solvents the unoccupied coordination sites of Mg can interact with polar residues of other Chl molecules, in particular the $>C=O$ group in the cyclopentanone ring (lower right in the Chl structure shown in Fig. 1.1). This is how dimers and multimers of anhydrous Chl probably are formed in dry nonpolar solvents.

Chl a at low concentration in a polar solvent such as ether, acetone, or pyridine is monomeric. In a "good" nonpolar solvent such as CCl_4 or benzene, Chl a exists mainly as dimers, even at low concentration. This has been confirmed by its molecular weight as revealed by osmometry of the solvent, and by infrared and NMR spectra. In "poor" nonpolar solvents (hydrocarbons), higher aggregates predominate.

Water is bound avidly by the Mg atom in Chl, and this alters the mechanism of dimerization and changes the properties of the monomeric and aggregated pigment in the foregoing solvents. The Q_y transition is shifted to greater wavelengths, and the chemical and ESR properties are altered. We shall see that the "special pair" chlorophylls in reaction centers appear to be hydrated.

Dry Chl *a* monomers show long wave (Q_y) absorption maxima ranging from 660 to 672 nm in various solvents. The dimer has two Q_y bands near 668 and 682 nm, and higher aggregates show bands spanning the range 663–700 nm, with relatively little of the 700 nm component. Intense absorption near 700 nm, as in the special pair of Photosystem 1, requires hydration of the Chl. These spectral properties can in principle account for the multiple spectral forms of Chl *a* in photosynthetic membranes, but the correlations are not decisive. Crystals of Chl *a* show an absorption maximum near 740 nm, not seen in healthy photosynthetic tissues.

The effects of dimerization on absorption and CD spectra of Bchl *a* in solution are shown in Fig. 7.7 (monomer in ether; dimer in CCl_4). The single monomeric absorption band at 770 nm is replaced in the dimer by one at 779 nm plus a shoulder near 815 nm. The integrated areas of the absorption bands are about the same for the monomer and the dimer. The small positive (left minus right) CD band at 775 nm for the monomer is replaced in the dimer by a much larger and more complex spectrum. The main feature is a large double band, but it does not have the simple symmetrical form of a pure dimer as described in the point dipole approximation. Bchl *a* crystals show broad absorption bands at wavelengths as great as 940 nm, but these bands do not resemble the bands seen in vivo. The combination of extreme shift (up to 890 nm) without broadening for the Q_y bands of Bchl *a* in vivo has no counterpart in vitro.

The spectrum of antenna Bchl *a* in membrane fragments from *Rhodopseudomonas sphaeroides* shows bands at 800 and 850 nm due to the B850 antenna component and at 875 nm due to B875 (recall Section 5.4). In preparations of isolated B850 the 800 nm band decays slowly with time or mistreatment (e.g., freezing and thawing, and illumination), apparently because one of the three Bchl molecules in the complex becomes dislodged. The stable 850 nm band is a property of the two Bchl molecules that remain bound to the protein; similarly, the 875 nm band is associated with the two Bchl molecules of the B875 complex.

We are not yet certain to what extent the large shifts of the long wave band of Bchl (and the smaller shifts for Chl *a*) in vivo are due to interactions of Bchl with other Bchl molecules (excitonic interactions) and to what extent they are due to interactions of Bchl with charged groups and inducible dipoles in the lipoprotein matrix. The spectra of Bchl in vitro would suggest that solvent effects cannot produce such large shifts, and that these shifts must be

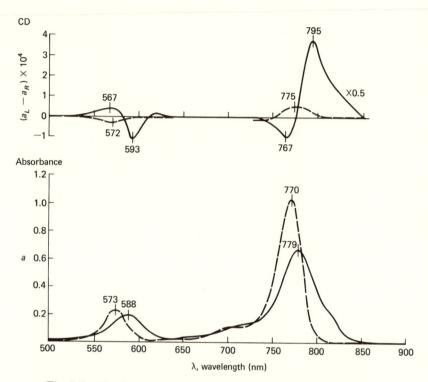

Fig. 7.7. Circular dichroism and absorption spectra of Bchl *a* as monomers in ether and as dimers in carbon tetrachloride. Solid curves: dimers, 1.1×10^{-4} M (on basis of monomer), optical path 1 mm. Dashed curves: monomers, 1.1×10^{-5} M, optical path 1 cm. Adapted from K. Sauer (1972), *Methods Enzymol.* *24*, 206–17 (Fig. 3).

due to Bchl–Bchl exciton interactions. However, we have found recently[8] that the antenna Bchl *a* of carotenoidless mutant *Rp. sphaeroides* can be brought to an apparently monomeric state in the membrane while retaining a strongly shifted (about 850 nm) absorption maximum. This antenna Bchl is destroyed by prolonged exposure to strong light. Under progressive photodestruction the absorption maximum shifts from about 860 to 853 nm. The progress of this shift, as a function of the fraction of pigment destroyed, fits a binomial statistical distribution as expected for individual targets (Bchl molecules) that are associated initially as dimers (two Bchl per antenna subunit) but become monomers when their dimer-partners are knocked out. The suspected conversion of dimers to monomers was confirmed by a progressive change in the CD spectrum: The initial dimerlike spectrum (as shown in Fig. 7.8) was lost in proportion to the computed loss of dimers.

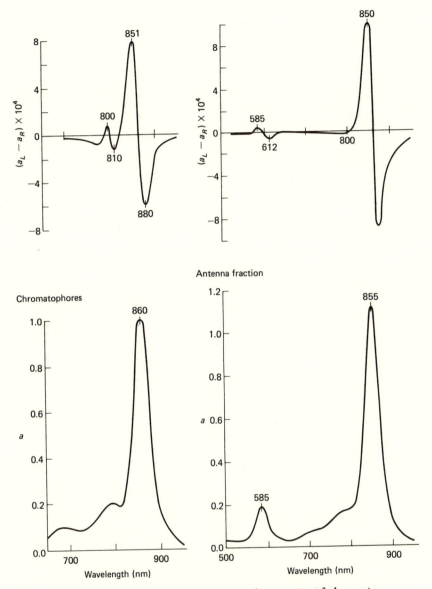

Fig. 7.8. Circular dichroism and absorption spectra of chromatophores (membrane fragments) of carotenoidless mutant *Rhodopseudomonas sphaeroides*, and of a purified antenna pigment–protein complex derived from these membranes. Adapted from K. Sauer and L. Austin (1978), *Biochemistry 17*, 2011–19 (Figs. 1 and 3).

It is reasonable to suppose that large shifts induced by the surroundings (not by Bchl–Bchl exciton interactions), to about 850 nm or beyond, are modified further in the antenna subunits by exciton interactions. These exciton interactions do not usually produce well-resolved absorption bands. They are indicated by resolved double CD bands (large positive and negative waves, as in Fig. 7.8) and in some cases by broadening of the absorption bands.

To summarize, a reasonable working hypothesis is that the major Q_y absorption bands of antenna Bchl in photosynthetic bacteria are composed of two or more nearly degenerate transitions, arising from exciton interactions but strongly shifted from the monomeric wavelength by other interactions of Bchl with its surroundings. In view of the composition of purified antenna fractions (two Bchl per subunit, plus a third labile Bchl in some cases), and the spectral similarity of these preparations to their counterparts in the living cell (Fig. 7.8), the major exciton interaction is probably dimeric within each subunit. The 800 nm absorption band of B850 is accompanied by only a weak single CD band comparable to that of a monomer. The labile Bchl responsible for the absorption and CD at 800 nm therefore appears to lack strong exciton interaction with the two that give the 850 nm band.

We do not know the relative orientations of the monomer Q_y transitions in B850 and other antenna pigment–protein complexes from bacteria, except that they are neither strictly parallel nor strictly perpendicular (for either of these orientations, Equation 7.1 gives a rotational strength of zero). Given the right relative translational coordinates, two monomers can have transition moments that are not orthogonal, interacting to produce dimer transitions of unequal amplitude with little or no difference in energy.

We shall now consider further evidence that there is strong dimeric exciton interaction between Bchl molecules in individual antenna subunits, and that weaker coupling exists between separate subunits.

If the exciton coupling between antenna subunits were as strong as that within them, the rate at which energy migrates to the reaction centers should be considerably greater than that which is observed. The lifetime of antenna fluorescence shows that about 100–500 psec are needed, on the average, for a quantum absorbed in the antenna to reach the reaction center. This is consistent with weak-to-moderate coupling as described in Section 2.5.

The absence of strong coupling between antenna subunits in vivo is also indicated by the fact mentioned earlier, that the absorption and CD spectra of antenna components in the membrane are changed very little (making due allowance for the effects of detergents) when these components are separated from the membrane and purified. This is shown in Fig. 7.8 for membrane fragments (chromatophores) and isolated antenna pigment–protein from carotenoidless mutant *Rp. sphaeroides*. The main difference in these spectra can be traced to the presence of reaction centers in the membrane fragments and not in the isolated antenna preparation. The reaction centers are responsible for the CD bands near 800 nm (see later, Fig. 7.9), missing in the antenna preparation. The shift of the main absorption band from 860 nm (membrane fragments) to 855 nm (isolated antenna) was caused by the detergent (Triton X-100) used for isolation. The absorption and CD spectra remained unchanged when the antenna fraction was dissociated into its component subunits by adding 1% sodium dodecyl sulfate.

The corresponding analysis of antenna components in green plant tissues is not as complete as in photosynthetic bacteria. The CD spectrum of the LH*a/b* component shows strong exciton interaction among Chl *b* molecules, but not among Chl *a* molecules. The CD of "Fraction 1" preparations shows strong exciton interactions among Chl *a* molecules absorbing near 680 nm.

Turning now to the reaction centers, let us first consider the nature of the photochemical electron donor, Chl or Bchl. In its oxidized form (P^+) this donor has an ESR spectrum characteristic of the cation radical of Chl or Bchl. In purple bacteria this ESR spectrum is not changed when the Bchl is modified by substituting ^{25}Mg for the common isotope ^{24}Mg. This shows that the photochemical oxidation of Bchl entails the removal of a π electron that interacts very little with the central Mg atom.

The ESR spectrum of the oxidized form of the photochemical electron donor also shows that it is generally not a single molecule but rather a "special pair" of Chl or Bchl molecules. In studies of chlorophyll radicals in vitro, the monomeric radicals, $Chl^+\cdot$ or $Bchl^+\cdot$, show an undifferentiated ESR band of width 9 gauss for $Chl^+\cdot$ and 13 gauss for $Bchl^+\cdot$. For anhydrous dimers and higher aggregates, the singly oxidized radical $(Chl)_n^+\cdot$ retains its band width of 9 gauss, but hydrated species of the type $(Chl\cdot H_2O)_n^+\cdot$ show much narrower bands (see Norris et al., suggested reading). Theory predicts a narrowing by the factor $\sqrt{n}$ if the unpaired electron is shared equally by all n molecules of Chl. Apparently the electron is delocalized in the hydrated multimer but not in the anhydrous one. The ESR band width of the oxidized electron donor in Photosystem 1 is about 7 gauss, close to $1/\sqrt{2}$ times the band width of the monomer. For the donor in Photosystem 2 the band width is about 8 gauss. It could be inferred that for Photosystem 1 the electron donor is a dimer of the type $(Chl\cdot H_2O\cdot Chl)^+\cdot$ or $(Chl\cdot H_2O)_2^+\cdot$ in its oxidized form, with the $+$ charge (or the unpaired electron) shared equally by both Chls. For Photosystem 2 the electron donor might be a monomer or a dimer; recent studies of Chl *a* in vitro[9] emphasize the idea that it is a monomer. In nonaqueous solvents the ESR band width of Chl $a^+\cdot$ ranges between 7.8 and 9.2 gauss depending on the solvent, and the midpoint potential of the couple $Chl/Chl^+\cdot$ can be as high as $+0.9$ V, appropriate for the oxidation of water in Photosystem 2. In the reaction centers of various photosynthetic bacteria the oxidized donor has band width about 9 to 9.5 gauss, again about $1/\sqrt{2}$ times the band width for monomeric $Bchl^+\cdot$. As a word of caution, the factor $1/\sqrt{2}$ can also arise if an unpaired electron is shared unequally by more than two molecules; for example, 1:4:1 in a set of three molecules. However, the implication that the donor is a hydrated dimer (special pair) of Chl or Bchl has been confirmed decisively in some instances by measurements of electron–nuclear double resonance (ENDOR) showing that the unpaired electron visits equally the π-system protons of two Chl or Bchl molecules. This has been done for the reaction centers of *Rp. sphaeroides* and Photosystem 1 of spinach, but not for Photosystem 2.

When bacterial reaction centers are poised at low redox potential so as to reduce the electron-accepting quinone, and then illuminated, a triplet state of Bchl is formed by a mechanism that will be discussed in Section 8.1. The zero field splitting parameters (Section 7.1) of the ESR spectrum of this triplet state show that the unpaired electron spins are confined to an elongated region approximating two Bchl molecules, and the absorbance changes attending triplet formation indicate that the triplet state is localized in the special pair that normally acts as photochemical electron donor. Other details of the zero field splitting have been interpreted to indicate that the angle between the normals to the two tetrapyrrole planes is about 48°. In fractions from green plants the triplet ESR spectrum is somewhat confusing, possibly because of contributions of antenna Chl *a* in its triplet state.

Spectra of absorption and light-induced absorbance changes of bacterial reaction centers were illustrated in Fig. 5.11 and discussed in Section 5.1. We shall now consider some details and interpretations of these spectra. Because there are strong exciton interactions among the molecules of Bchl and Bpheo in each reaction center, as shown most clearly by their circular dichroism, one must be prepared to regard the entire set of absorption bands as the communal property of the six chromophores. There is heuristic value, however, in a set of simplifying assumptions whereby some bands are identified primarily with one or more specific molecules in the reaction center. In this spirit we can consider the following identifications.

The band at 535 nm, resolved at low temperature into two at 532 and 547 nm, represents the Q_x transitions of the two molecules of Bpheo. The absence of strong CD in this region indicates that these transitions have little exciton interaction. This interpretation is supported by the observation that the 547 nm band is bleached, and the 532 nm band remains unperturbed, when one molecule of Bpheo is reduced in the earliest photochemical step (Reaction 5.1, and see Section 8.1).

The band near 600 nm contains the unresolved Q_x bands of the four molecules of Bchl. Some resolution is evident in the light-induced change that reflects oxidation of P870, the special pair (Fig. 5.11*b*).

The band at 757 nm can be identified mainly with the two molecules of Bpheo. This Q_y band is nearly identical to that of monomeric Bpheo in solution. Some exciton interaction, either between the two Bpheos or between Bpheo and Bchl, is indicated by a small double CD effect around 760 nm, the positive branch hidden by the larger double CD band centered near 800 nm (Fig. 7.9). The CD effect around 760 nm is scarcely perturbed

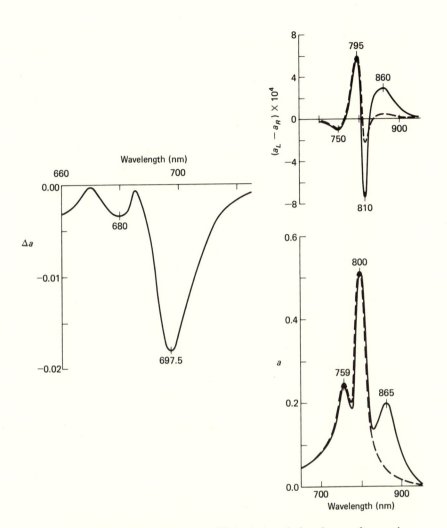

Fig. 7.9. Left: A spectrum of light-induced absorbance changes in a fraction from spinach enriched for Photosystem 1, reflecting the oxidation of the "P700" Chl *a* complex. Adapted from K. D. Philipson, V. L. Sato, and K. Sauer (1972), *Biochemistry 11*, 4591–5 (Fig. 2). Right: Absorption and CD spectra of reaction centers isolated from carotenoidless mutant *Rp. sphaeroides*, measured in weak light (solid curves) and in light strong enough to oxidize the Bchl special pair (dashed curves). Adapted from K. Sauer and L. Austin (1978), *Biochemistry 17*, 2011–19 (Fig. 2).

when the special pair is oxidized, indicating little or no exciton interaction between Bpheo and the special pair. The perturbation that does exist can be ascribed to a shift of the 757 nm band to greater wavelengths,[10] seen more clearly in Fig. 5.11 than in Fig. 7.9. The 757 nm band loses about half of its intensity when one Bpheo becomes reduced in the first photochemical step (Section 8.1).

The most conspicuous consequence of oxidation of the special pair is complete loss of the 870 nm band, with the 803 nm band remaining intact except for a small blue-shift. This suggests that for the 870 nm transition, the electron redistribution is confined mainly to the special pair, whereas the 803 nm band is mainly a property of the "voyeur" Bchls.

The changes of absorbance around 800 nm (Fig. 5.11*b*) were ascribed initially to a blue-shift of the 803 nm band attending oxidation of the special pair, but this is only part of the story. The polarizations of the positive and negative branches of this "shift" are not the same (see later in this section); it thus cannot represent merely the shift of the absorption band of a single transition. At low temperature the 803 nm band has a distinct shoulder at about 810 nm, and this shoulder disappears along with the 870 nm band when the special pair is oxidized. At the same time, a new band appears at about 795 nm. Oxidation of P870 thus causes the loss of bands near 810 and 870 nm, and the gain of a band near 795 nm. Taking these effects into account, less than half of the light-induced "wave" around 800 nm can be attributed to blue-shift of the 803 nm (voyeur Bchl) band.

One simple interpretation of these absorbance changes is that the pair of bands at 810 and 870 nm reflects dimeric association of the special pair, the resultant of monomeric Q_y transitions. The singly oxidized dimer $B_2^+\cdot$, with the + charge resonating between two molecules of Bchl, has aspects of a monomerlike spectrum with a single band at 795 nm. The changes near 600 nm (Fig. 5.11*b*) also suggest, although less strikingly, the loss of dimeric absorption bands at 605 and 630 nm and the gain of a monomerlike band near 575 nm. Because the 870 nm band is far more intense than the 810 nm band, the monomeric Q_y transitions must be nearly parallel and predominantly collinear (Fig. 7.3*e*); by similar reasoning the monomeric Q_x transitions that yield the dimeric bands at 605 and 630 nm are nearly parallel and predominantly side by side (Fig. 7.3*f*).

The CD spectra shown in Fig. 7.9 are qualitatively consistent with this view. Oxidation of the special pair causes principally the loss of dimerlike CD bands, negative at 810 nm and positive at 860 nm. The features that remain when the special pair has been oxidized can be attributed to the Bpheo (negative CD below and positive above 760 nm), the voyeur Bchl (positive below and negative above 800 nm), and perhaps a weak positive band for the "monomeric" 795 nm absorption band of $B_2^+\cdot$.

A contrasting interpretation is that the 870 nm band results from dimeric association of strictly parallel monomeric Q_y transitions, and similarly for the

605 nm band (parallel Q_x transitions). The 810 and 630 nm bands are then secondary consequences of the special pair interacting with one or more voyeur Bchls.

Both of these views are consistent with the observation that the orientations of transitions (relative to axes fixed in the reaction center) are constant over the entire 870 nm band, and also over the entire 605 nm band. This allows us to suppose that each of these bands is due to a single electronic transition in a dimer, the resultant of monomeric transitions. Neither view is compatible with the large positive CD band at 870 nm, much larger than the negative CD band at 810 nm. The CD at 870 nm could, however, be due partly to chirality imposed on the special pair by the protein ("solvent effects").

But both of these interpretations, whatever their heuristic value, are naïve. With four interacting chromophores (perhaps six, if we count the Bpheos), the 870 nm band could well encompass parallel but distinct transitions of the complex (this could not arise in a dimer).

With complications such as multiple chromophore interactions, extended dipoles, and solvent effects we should not expect an early definitive resolution of this complex problem.

The reaction center of *Rp. viridis*, containing Bchl *b* and Bpheo *b*, has optical counterparts of all the features just described for *Rp. sphaeroides*, but at greater wavelengths. The same interpretive arguments apply to both organisms. *Rp. viridis* has the advantage that the spectral features in the near infrared are spread over a greater range of wavelengths and are easier to resolve.

The reaction center of Photosystem 1, studied in fractions enriched for this reaction center, shows optical features somewhat like those of the bacterial reaction center[11] (Fig. 7.9). Oxidation of the special pair (known popularly as P700) in Photosystem 1 causes absorbance changes that can be dissected into a loss of bands at 680 and 698 nm (dimer bands?) and the gain of a band at 686 nm (monomerlike band of the oxidized special pair?). The light-induced change of CD, not shown, has two poorly resolved sets of positive and negative components that underlie a much larger CD of antenna Chl *a* in the preparation. The absorption bands of P700 in Fig. 7.9 are at shorter wavelengths than in vivo, probably because of exposure to Triton X-100.

Let us turn now to studies that bear on the orientations of transition moments of antenna and reaction center pigments in relation to each other and to the plane of the photosynthetic membrane.

If a randomly oriented collection of thylakoid membranes (a suspension of algae, chloroplasts, or chloroplast fragments) is exposed to polarized light of wavelength less than about 680 nm, the resulting fluorescence collected from the sample shows little polarization. This depolarization is due to energy transfer from the photoselected absorbing oscillators to randomly aligned emitting ones. On the other hand, an oriented preparation of membranes shows polarized fluorescence, with the electric vector roughly parallel to the plane of the membrane, whether the actinic light is polarized or not. This shows that the emitting oscillators, the long wave transitions of the fluorescent antenna Chl *a* associated with Photosystem 2, lie approximately in the plane of the membrane. This was first shown by R. A. Olson et al. (see the suggested reading) by an ingenious use of a microscope in conjunction with an infrared image converter. Cells of the green alga *Euglena*, containing chloroplasts readily visible in the microscope, were exposed to polarized or unpolarized blue light. A filter was interposed in the viewing path to transmit only the fluorescent light, and the image formed by this light was viewed through the image converter so that it could be seen. With a polarizer placed in the viewing path it could be seen that the electric vector of the fluorescence was parallel to the planes of the membranes. In a companion experiment the light transmitted by the chloroplasts was viewed with the electric vector parallel or perpendicular to the membranes. Light of wavelengths greater than about 700 nm was polarized parallel to the membranes; light below about 660 nm showed no obvious polarization. Between 660 and 770 nm the membranes were too opaque to be viewed edgewise.

There have been several reports of linear dichroism in chloroplasts and thylakoid membrane fragments oriented in various ways: by photoselection (measuring fluorescence or the bleaching of P700), by laminar flow, by drying onto plates (in some cases with gentle brushing during drying), and by a magnetic field. Anisotropic diagmagnetism of the membranes, of unknown origin, aligns the membranes with their planes perpendicular to the magnetic field vector. The consensus of these experiments is as follows:

Long wave Q_y transitions of antenna Chl *a* ($\geqslant 680$ nm) lie within about 20° of the plane of the membrane. P700 and P680 (the long wave Q_y transitions of the special pairs of Photosystems 1 and 2) also lie nearly parallel to the plane of the membrane, as do the 810–825 nm transitions of these special pairs in their oxidized forms.[12] There is disagreement about the orientations of the Q_x transitions of these Chls. In principle these molecules could be nearly coplanar, with the planes probably tilted somewhat from the plane of the membrane. This could facilitate energy transfer to the reaction centers. Carotenoids lie with their long axes nearly parallel to the membrane planes.

There is little evidence for long-range molecular order with respect to different directions in the plane of the membrane.

Polarized light has also provided fragmentary indications of the effects of divalent cations in altering the distribution of quanta between the two photosystems in green plant tissues. To cite a single example, J. Biggins and J. Svejkovsky[13] have studied the effects of divalent cations on the linear dichroism of chloroplasts oriented by a magnetic field. They found that a component of Chl *a* absorbing at 690 nm and associated with Photosystem 1 became reoriented by the addition of Ca^{2+} or Mg^{2+} (about 3 mM). Addition of the cation caused the Q_y transition at 690 nm to become aligned more closely with the plane of the thylakoid membrane. This effect was absent in chloroplasts whose internal structure had been fixed by cross-linking the proteins with glutaraldehyde. It is hoped that a greater variety of such experiments will help to elucidate the mechanisms by which cations regulate the quantum distribution.

Studies of linear dichroism and photoselection have reached their greatest elaboration with cells, membrane fragments, and isolated reaction centers of photosynthetic bacteria. The information about structure gleaned from these studies compensates a little for the absence of X-ray crystallographic analysis; no one has yet succeeded in crystallizing hydrophobic proteins such as reaction centers.[14]

Several methods have proved useful for orienting membranes and reaction centers of photosynthetic bacteria. Orientation of whole cells is achieved by a magnetic field if the membranes are arranged in a regular way in the cells: as stacks or concentric cylindrical layers in *Rp. viridis* and *Rp. palustris*. The cells become aligned with the magnetic vector perpendicular to the greatest amount of membrane surface. Membrane fragments tend to lie flat when dried onto glass plates. These membrane fragments can also be treated so as to destroy the antenna Bchl selectively, exposing the absorption spectrum of the reaction centers in an oriented context. Isolated reaction centers can be dried into films with gelatin[15] or polyvinyl alcohol. If the films are then humidified, stretched, and dried again, an axis of symmetry in the reaction centers (or perhaps in a linear aggregate of reaction centers) becomes aligned with the direction of stretching. In these stretched films the reaction centers are disposed randomly with respect to rotation about the axis of symmetry. A similar orientation is achieved if reaction centers or membrane fragments are embedded in polyacrylamide gels, and the gels are then squeezed so as to elongate them.

By studying the linear dichroism of absorbance and light-induced absorbance changes in these variously oriented preparations, one can obtain the angles that a variety of optical transitions (in Bchl, carotenoids, cytochromes, etc.) make with an axis such as the normal to the membrane or the direction of stretching in a film. By the technique of photoselection, measuring light-induced changes of absorbance in cells, membrane fragments, and pigment–protein complexes such as reaction centers, one can obtain the angles between the pairs of transition moments involved in excitation and measurement. Finally, by combining the results from linear dichroism and photoselection measurements, one can compute both polar coordinates for a transition: the declination from the axis normal to the membrane, and the azimuthal coordinate around that axis (using one transition to establish an arbitrary zero for the azimuthal coordinate).[16] This has been done for several transitions of reaction-center pigments in the chromatophore membrane.[17] For a flat piece of membrane containing reaction centers, the declination from the normal is approximately the same throughout the membrane for a given transition. The azimuthal coordinates of transitions in the reaction centers are local:

They have definite values with respect to an arbitrary zero in each reaction center, but there is no long-range coherence throughout the membrane.

The principal results of these studies will now be summarized.

In photoselection studies with reaction centers from *Rp. sphaeroides*, measuring the bleaching of the 870 nm absorption band or the corresponding fluorescence at 900 nm, the polarization is close to the maximum theoretical value $P = \frac{1}{2}$ that signifies a single transition (or two distinct but parallel transitions) and not a pair of degenerate transitions. The single CD band coincident with the 870 nm absorption band (Fig. 7.9) supports this conclusion. The counterpart in reaction centers from *Rp. viridis*, the 980 nm band, has the same properties.

The Q_x band near 600 nm in reaction centers from *Rp. sphaeroides* is complex, but the Q_x components that show reversible bleaching due to oxidation of the special pair (Fig. 5.11*b*) appear to be due to single transitions centered at 605 and 630 nm, perpendicular to each other. The 605 nm transition of the special pair is also perpendicular to the 870 nm (Q_y) transition. These orthogonal relationships are expected for the transitions in a dimer of Bchl. The Q_x transitions of the two Bpheo molecules, at 532 and 547 nm, are nearly orthogonal. This is not required by theory, as these transitions are not exciton-coupled. The angle between the 760 nm Q_y transitions of the Bpheos, resolved in refined photoselection studies by A. Vermeglio et al.,[18] is 56°. The average orientation of these 760 nm transitions is perpendicular to the 870 nm transition of the special pair. The transitions around 800 nm are not well resolved, but the 810 nm component associated with the special pair makes a large angle (perhaps 90°) with the 870 nm transition. The monomer-like 795 nm component of the oxidized special pair is approximately parallel to the 870 nm transition, as is the 1250 nm component.

For the reaction center in its native context, the "special pair" transitions at 870 and 630 nm and those of the oxidized special pair at 795 and 1250 nm all lie close to the plane of the membrane. The special pair transitions at 810 and 605 nm, and the Q_x transitions of Bpheo, and one of the Q_y transitions of Bpheo, all point out of the membrane, making angles ranging from about 30° to 70° with the plane of the membrane. One Q_y transition of Bpheo lies close to the plane of the membrane.

The Q_y transition of Bpheo that points out of the membrane plane probably belongs to the photochemically active Bpheo molecule. In studies of linear dichroism of *Rp. viridis*, reduction of the active Bpheo causes partial bleaching of a band at 790–800 nm, the counterpart of the 760 nm Bpheo band in *Rp. sphaeroides*. This bleachable transition is roughly perpendicular to the membrane. The corresponding Q_x transition at 547 nm lies closer to the plane of the membrane.

The axis of symmetry around which isolated reaction centers are aligned in stretched gelatin films lies, in vivo, in the plane of the membrane, parallel to the 630 nm transition of the special pair.

For the antenna pigments in photosynthetic bacteria the carotenoids have their long axes about 45° out of the plane of the membrane, the average of the Q_y transitions of Bchl lies close to the plane, and the average Q_x of Bchl points out of the plane. For *Rp. sphaeroides* the Q_y transitions of both B850 and B875 lie in the plane of the membrane, or close to it. The labile 800 nm-absorbing molecule of Bchl in B850 has its Q_x transition also nearly parallel to the membrane, as can be seen by the dichroism of the bleaching at 590 nm when this molecule is disengaged and denatured. The Q_x (590 nm) transitions of the remaining antenna Bchl are approximately perpendicular to the membrane. Therefore, since the X- and Y-polarized transitions define the average tetrapyrrole plane, this plane is nearly parallel to the membrane for the 800 nm–absorbing Bchl and perpendicular to the membrane for the other Bchls in B850 and B875.

One might have hoped that these detailed studies would reveal nice geometrical relationships between the orientations of Bpheo and the Bchl special pair in reaction centers, and between the reaction-center pigments and antenna pigments – that something especially striking would leap to our delighted eyes. So far this has not happened. It can only be said that the antenna and reaction-center components could be coplanar in ways that could facilitate the transfer of excitation quanta to the reaction centers and the transfer of electrons in them.

In view of the rapidity and high quantum efficiency of the photochemistry (Section 8.1), one might also imagine that the reaction-center protein serves to hold the pigments in exquisitely precise relation to each other. This is contraindicated by the recent synthesis in vitro of a model reaction center[19] that has no protein and that approximates the kinetics of photochemistry in natural reaction centers. The main function of the protein seems to be to anchor the reaction centers in an orientation that gives directionality to the transfer of electrons (and subsequently protons) across the membrane.

Linear dichroism is beginning to be studied by ESR, to obtain the orientations of microwave transitions of electron spins in radicals of the cytochromes that act as secondary electron donors in general and the iron–sulfur centers that act as electron acceptors in Photosystem 1. Such studies can also be applied to quinone radicals and to the Chl and Bchl triplet states. The results are fragmentary, and definitive statements about them seem premature.

Further research with polarized light can proceed profitably along several lines. In bacterial reaction centers the complexity of transitions in the region 780-820 nm needs to be unraveled with better wavelength resolution and variation of parameters such as temperature. Even if we achieve detailed knowledge of the angular coordinates of the reaction-center pigments, we shall still need relative translational coordinates in order to assemble a structure for the reaction center. This is being approached in several ways: (1) by interpreting reaction kinetics in terms of new theories of vibronically assisted quantum mechanical electron tunneling; (2) by observing the natures of magnetic interactions between pigment molecules and their surroundings (especially other chlorophylls, pheophytins, quinones, and Fe^{2+}) as manifested in ESR spectra; and (3) by estimating the strengths of electric dipole interactions among the pigments. The chromophores appear to be of the order of 1.0–2.0 nm from each other, but relative positions are uncertain. Perhaps this problem can be worked out by analyzing the pigment band shifts that seem to be caused by local electric fields in the reaction centers in various states of oxidation and reduction.

More about structures of photosynthetic components can be learned by extending the new refinements combining optical and magnetic resonance spectrometry, and those optical techniques (infrared and Raman spectroscopy) that give information about vibrations and rotations of specific atomic groups.

All of these techniques, it is hoped, can give us more structural information about green plant and algal photosynthetic membranes and their components. Our knowledge in this area is in a relatively primitive state.

Finally, the technique of photoselection can be used to estimate the sizes of reaction-center particles and other membrane fractions in aqueous media, by observing the rate at which a photoselected sample becomes randomized through rotation. The same technique, measuring the rate of relaxation of photo-dichroism, can be used to probe the rotational mobilities of molecules such as quinones and cytochromes in the membrane. The reaction centers of *Rp. sphaeroides* do not tumble appreciably in the membrane over periods of many seconds.

SUGGESTED READINGS

Abu-Shumways, A., and Duffield, J. J. (1966). Circular dichroism – Theory and instrumentation. *Anal. Chem. 38*, 29A–58A.

Tinoco, I., Jr. (1963). The exciton contribution to the optical rotation of polymers. *Radiation Res. 20*, 133–9.

Sauer, K. (1974). Primary events and the trapping of energy. In *Bioenergetics of Photosynthesis*, Govindjee, ed., pp. 115–81 (especially Section 2, Chlorophyll spectroscopy, and Section 3, Chlorophylls in photosynthetic membranes). Academic Press, New York.

Hofrichter, J., and Eaton, W. A. (1976). Linear dichroism of biological chromophores. *Annu. Rev. Biochem. Bioeng. 5*, 511–60.

Albrecht, A. C. (1961). Polarizations and assignments of transitions: The method of photoselection. *J. Mol. Spectrosc. 6*, 84–108.

Swartz, H. M., Bolton, J. R., and Borg, D. C. (1972). *Biological Applications of Electron Spin Resonance*. Wiley-Interscience, New York.

Feher, G. (1970). *Electron Paramagnetic Resonance with Applications to Selected Problems in Biology*. Gordon and Breach, New York.

Clarke, R. H., Connors, R. E., and Frank, H. A. (1976). Investigation of the structure of the reaction centre in photosynthetic bacteria by optical detection of triplet state magnetic resonance. *Biochem. Biophys. Res. Commun. 71*, 671-5.

Leigh, J. S., Jr. (1978). EPR studies of primary events in bacterial photosynthesis. In *The Photosynthetic Bacteria*, R. K. Clayton and W. R. Sistrom, eds., Chapter 23, pp. 431-8. Plenum, New York.

Ballschmiter, K., Truesdell, K., and Katz, J. J. (1969). Aggregation of chlorophyll in nonpolar solvents from molecular weight measurements. *Biochim. Biophys. Acta 184*, 604-13.

Dratz, E. A., Schultz, A. J., and Sauer, K. (1966). Chlorophyll–chlorophyll interactions. *Brookhaven Symp. Biol. 19*, 303-18.

Norris, J. R., Uphaus, R. A., Crespi, H. L., and Katz, J. J. (1971). Electron spin resonance of chlorophyll and the origin of Signal 1 in photosynthesis. *Proc. Natl. Acad. Sci. U.S. 68*, 625-8.

Norris, J. R., and Katz, J. J. (1978). Oxidized bacteriochlorophyll as photoproduct. In *The Photosynthetic Bacteria*, R. K. Clayton and W. R. Sistrom, eds., Chapter 21, pp. 397-418. Plenum, New York.

Goedheer, J. C. (1957). Optical properties and in vivo orientation of photosynthetic pigments. Doctoral thesis, Netherlands State University, Utrecht.

Olson, R. A., Butler, W. L., and Jennings, W. H. (1961). The orientation of chlorophyll molecules in vivo: Evidence from polarized fluorescence. *Biochim. Biophys. Acta 54*, 615-17.

Olson, R. A., Butler, W. L., and Jennings, W. H. (1961). The orientation of chlorophyll molecules in vivo: Further evidence from dichroism. *Biochim. Biophys. Acta 58*, 144-6.

Breton, J., and Roux, E. (1971). Chlorophylls and carotenoids states in vivo. I–A linear dichroism study of pigments orientation in spinach chloroplasts. *Biochem. Biophys. Res. Commun. 45*, 557-63.

Part III

PHOTOCHEMICAL CHARGE SEPARATION, SECONDARY TRANSPORT OF ELECTRONS AND PROTONS, AND OXYGEN EVOLUTION

We turn now to the details of photochemical charge separation in the reaction centers of bacteria and plants, to the sequences of electron transport that follow from the primary charge separation, and to the attendant transport of protons across the photosynthetic membranes. In bacteria (except for the green sulfur bacteria, about which we know little) the pattern of electron transport is predominantly cyclic, but with paths that provide for the exchange of electrons and H^+ with external substrates. In plants the main path is from H_2O, through Photosystems 2 and 1, ending at NADP. The involvement of H transfer in carbon metabolism will not be considered here. A cyclic flow of electrons driven by Photosystem 1 has been delineated, but its quantitative importance remains uncertain. The net translocation of H^+ coupled to electron transport is from outside to inside the thylakoids in chloroplasts, and from the cytoplasm to the periplasm (from inside to outside the cell) in bacteria. The latter direction is topologically the same as "outside to inside" for vesicles bounded by intracytoplasmic membranes (see the introduction to Part II). This translocation of protons generates electrochemical gradients across the membrane; these gradients and their relation to ATP formation will be the main subject of Part IV.

For the most part the carriers of electrons and protons are well known enzymes and coenzymes, notably cytochromes, quinones, and ferredoxins (protein-bound Fe—S centers). It is only in the path from H_2O to the reaction center of Photosystem 2 that we have almost total ignorance of the chemical natures of the functional components (we do know that manganese is involved somehow). Our fragmentary understanding of the mechanism of oxygen evolution is based largely on phenomenology involving light, oxygen, pH changes, and the actions of a few inhibitors and artificial electron carriers.

Our knowledge of photochemistry, electron transport, and H^+ involvement in photosynthesis has come mainly from the observa-

165

tion of light-induced (and chemically induced) changes of absorbed and emitted light and ESR; usually these changes signal the oxidation or reduction of molecules. These observable phenomena, in the form of spectra and time dependence, have served to identify the reactions of specific substances and to reveal the dynamics of their interactions. The use of brief, intense flashes of actinic light (single, double, and multiple flashes) has been especially informative in showing the kinetics and the networks of interaction among the functional entities.

The basic observations have been extended by varying such parameters as temperature, ambient redox potential, and pH; this gives information about reaction mechanisms and chemical properties of the reactive components. Sequences of electron carriers have been identified through various deletions and additions of substances: deletion in mutants, or by the application of specific inhibitors, or by selective extraction; addition of artificial electron donors and acceptors, and replacement of deleted components or analogous substances. On the whole, electron-transport sequences inferred by these methods have made sense thermodynamically. Following the light reaction, the flow of electrons is downhill in energy, through carriers of ever more positive oxidation–reduction potential. This downhill flow of electrons is nicely tuned to the need for stabilization against back-reactions and is coordinated with a transduction from redox energy to the energy of ionic gradients across membranes. Only rarely does a single step in electron transfer seem to waste an inordinate amount of energy by crossing a large gap in redox potential.

Our knowledge of events in reaction centers is far more advanced for the purple bacteria than for the green plants, but the insights gained by studying the isolated bacterial reaction centers have pointed the way to a rapidly accelerating understanding of the reaction centers of Photosystems 1 and 2.

The topics of photochemistry and related electron flow and H^+ transport comprise the bulk of most contemporary conferences on photosynthesis. The content of Part III can therefore be found in elaborate detail in many recent reviews and accounts of symposia. In coordinating this material I shall try to be didactic and brief, at the risk of being cryptic.

8 Reaction centers: photochemical charge separation and interaction with nearest electron donors and acceptors

8.1 Charge separation in reaction centers of photosynthetic bacteria

The first direct evidence of photochemical charge separation in reaction centers came with Duysens's observation of the reversible light-induced bleaching of P870 in purple bacteria. B. Kok and H. T. Witt then discovered independently the reversible bleaching of P700 in chloroplasts, and Witt and collaborators later described the reversible bleaching of P680, the primary electron donor in Photosystem 2. The detection of P680 was more difficult because the oxidized form P^+ has a very short lifetime, owing to rapid reduction by secondary electron donors (see Section 8.2).

The absorbance changes attending oxidation of P870 have been described in Section 5.1 (Fig. 5.11b) and interpreted in Section 7.2. The evidence, principally from ESR and ENDOR spectra, that P870 is a special pair of Bchl a molecules in the reaction center was given in Section 7.2. Measurements of absorbance changes following brief laser flashes have shown that $P870^+$ is formed within about 3 psec.

The best evidence that $P \longrightarrow P^+$ involves the transfer of a single electron (from P to an acceptor) comes from redox titrations of the couple P/P^+. The midpoint potential of this couple has been measured by exposing reaction centers to various ambient redox potentials (using mixtures of ferri- and ferrocyanide), and determining the fractions of P870 in the forms P and P^+ by direct measurement of absorbance at 870 nm or by the magnitude of a light-induced bleaching of the residual P. For purple bacteria these redox titrations have given $E_m = 0.4$ to 0.5 V, independent of pH, with $n = 1$ (from Equation 4.4). The value $n = 1$ shows that P and P^+ differ by just one electron. Accepting this conclusion, the extinction coefficient ϵ of P870, as given in the ordinate of Fig. 5.11, could be determined by measuring a coupled reaction between P^+ and mammalian cytochrome c:

$$Cyt_{red} + P \xrightarrow{\text{light}} Cyt_{red} + P^+ \longrightarrow Cyt_{ox} + P$$

(see Reaction 5.2). The oxidation of cytochrome and the reduction of P^+ were monitored by the absorbance changes at 550 and 870 nm, respectively. The molar extinction coefficient for cytochrome oxidation is well known, and given that 1 mole of P^+ is reduced for each mole of cytochrome oxidized, the extinction coefficient of P could be computed.

Typical parameters of absorbance changes, ESR spectra, and E_m are tabulated (Table 8.1) for purple bacteria that contain Bchl a or Bchl b, and for the green sulfur bacteria, in which the antenna is Bchl c, but the reaction centers contain Bchl a and no Bchl c. The ESR band widths show the narrowing (compared with monomeric Bchl) appropriate for a dimer, except in the case of *Rp. viridis*, where the band is anomalously broad (narrower than the monomeric band, but broader than the expected width of $13/\sqrt{2}$ or 9.2 gauss for the dimer). Because the optical properties suggest a special pair in this case also, we may wonder whether P980 in *Rp. viridis* has anomalous sources of ESR broadening, or whether the $+$ charge in P^+ is shared, but not equally, by the two Bchl molecules.

The reaction centers of green sulfur bacteria differ markedly from those of purple bacteria in their optical and redox properties; we shall return to these differences. The photochemical electron donor, nevertheless, appears to be a special pair of Bchl a molecules.

The photochemical oxidation of P870 (or P890, etc.) of course implies the existence of a partner, the electron acceptor. The question of identities and properties of "primary" electron acceptors has been relatively difficult to resolve, mainly because the light-induced changes of optical absorption and ESR spectra have been either smaller or broader than those of P/P^+, and hence harder to detect and monitor. In this context the word "primary" has several times been a misleading cliché. For years the primary acceptor in purple bacteria was thought to be a quinone, or a Q–Fe complex, but we now regard Bpheo as an earlier acceptor, and there is evidence that the voyeur Bchls may play a still earlier role in the transfer of an electron from the special pair. Our recognition of ever more "primary" acceptors has paralleled our ability to resolve time in optical and ESR measurements. When the time resolution for measuring changes of optical absorbance was pushed from 10^{-6} sec to 10^{-12} sec, the role of Bpheo (and possibly Bchl) as an electron acceptor emerged. Will we see anything new between 10^{-12} and 10^{-15} sec? We need not seek further than that; 10^{-15} sec is less than the time needed for one oscillation of the light wave in the visible or near infrared.

The word "primary," qualified by quotation marks, continues to be applied in the scientific literature to the quinone, with "intermediary acceptor" (designated I) applied to the Bpheo-voyeur Bchl complex. A similar semantic confusion is found in published descriptions of electron acceptors in Photosystem 1.

We can begin our account of photochemical electron acceptors in purple bacteria with a scheme showing the movements of an electron after the Bchl special pair has been excited by light:

$$P \xrightarrow{\sim 3\ psec} I \xrightarrow{\sim 200\ psec} Q_a \xrightarrow{\sim 0.1\ msec} Q_b \longrightarrow \text{Quinone pool}$$

$$(8.1)$$

Table 8.1. *Properties of photochemical electron donors in three species of photosynthetic bacteria*

Organism	Electron donor (special pair)	Peak wavelengths (nm) and signs of light-absorbance changes	E_m (V)	ESR	
				g Value	Band width (gauss)
Rhodopseudomonas sphaeroides[a]	P870 (Bchl *a*)	−810, −870 +795, +1250	0.45 to 0.5	2.0026	9 to 9.5
Rp. viridis	P980 (Bchl *b*)	−850, −980[b] +810, +1370	0.4 to 0.5	2.0026	11 to 12
Chlorobium sp. (green sulfur bacteria)	P840 (Bchl *a*)	−830, −842 +790, +1160	0.25 to 0.35	2.003	9.2

[a] Also typical of other purple bacteria that contain Bchl *a*.
[b] 960 nm in isolated reaction centers.

or, in alternative terminology,

B_2 (special pair) $\longrightarrow$ Bpheo (and voyeur Bchl?) $\longrightarrow Q_a \longrightarrow Q_b \longrightarrow$ Q pool

Optical evidence that UQ is an electron acceptor (Q_a) in reaction centers of *Rp. sphaeroides*, shown in Fig. 5.11c, was developed about 1970. The absorbance changes reflecting UQ $\longrightarrow$ UQ$^-$· are accompanied by shifts of the absorption bands of Bpheo; these have nothing to do with the role of Bpheo as an electron acceptor. Smaller shifts of some bands of Bchl also attend the reduction of Q_a. The pattern of these shifts changes when the electron moves from Q_a^- to Q_b, providing a way to monitor this secondary electron transfer.

Another way to monitor $Q_a^- Q_b \longrightarrow Q_a Q_b^-$ is by the use of closely spaced double flashes. In *Chromatium vinosum*, one of a pair of bound cytochromes can re-reduce P$^+$, after a flash, in 1-2 μsec:

$$\frac{Cyt(red)}{Cyt(red)} \cdot PIQ_a Q_b \xrightarrow{h\nu} (P^*IQ_a Q_b \longrightarrow P^+I^-Q_a Q_b \longrightarrow$$

$$P^+IQ_a^-Q_b) \xrightarrow{1-2\ \mu sec} \frac{Cyt(ox)}{Cyt(red)} \cdot PIQ_a^-Q_b$$

A second flash, given before Q_a^- has delivered its electron to Q_b, does not oxidize the second cytochrome; the state P$^+$I$^-$ induced by the second flash collapses to PI before this can happen. The second cytochrome can be oxidized only after Q_a^- has passed its electron to Q_b and is again ready to take an electron from I$^-$, thereby preventing the back-reaction between P$^+$ and I$^-$.

By these methods the half-time for $Q_a^- Q_b \longrightarrow Q_a Q_b^-$ has been found to be about 200 μsec in *Rp. sphaeroides* and about 60 μsec in *Chr. vinosum*. The speed of this reaction, and also the speed of electron transfer from cytochrome to P$^+$ (about 1 to 100 μsec in various species of purple bacteria), prevents a wasteful back-reaction between P$^+$ and Q_a^-.

Q_a receives an electron from I$^-$ in about 200 psec after a flash that forms P$^+$I^-Q_a, judging from the decay of absorbance changes associated with I$^-$. In the state P$^+$IQ$_a^-$, a back-reaction (P$^+$IQ$_a^- \longrightarrow$ PIQ$_a$) proceeds with $t_{1/2}$ about 60-80 msec at room temperature and about 20 msec below 200 K (somewhat faster in *Rp. viridis*). We do not know why this back-reaction is faster at low temperature; the mechanism probably involves tunneling.[1] If the electron has moved on to Q_b or to a secondary pool of UQ, the back-reaction is much slower, of the order of 1 sec or more at room temperature. Below about 220 K there is no appreciable flow of electrons from Q_a^- to Q_b, or from secondary quinones to P$^+$.

Under physiological conditions Q_a never becomes doubly reduced. If the reaction center is excited while in the state PIQ$_a^-$, the product P$^+$I$^-Q_a^-$ reverts to PIQ$_a^-$ far more rapidly than it goes to P$^+$IQ$_a^{2-}$. Q_b can be doubly reduced; details of the interactions between the "one-electron" chemistry of Q_a and the "two-electron" chemistry of Q_b will be considered in Section 9.3.

The quinones associated with bacterial reaction centers can be removed and replaced selectively by suitable methods of extraction, a characteristic that has been of great value in exploring the interactions and identities of these quinones, and in showing their obligate roles in photosynthesis. Quinones can be removed from reaction centers by suitable exposure to detergents combined with *o*-phenanthroline,[2] or by extraction with hexane, with or without a trace of methanol (hexane removes secondary quinones from dried reaction centers; with methanol, Q_a is also removed). By these means one can prepare reaction centers with a complete complement of quinones including a secondary pool, or with just Q_a and Q_b, or with Q_a alone, or with no quinone at all. Also one can replace the native quinones with chemical analogues. These techniques have shown that Q_a is essential in stabilizing the primary charge separation, that Q_a is ubiquinone (UQ) in most bacteria but is menaquinone (similar to vitamin K) in *Chr. vinosum* and in *Rp. viridis*, that the secondary quinones are UQ in all purple bacteria (as far as is known), and that a variety of quinone analogues can be substituted for Q_a with only minor changes of reaction kinetics.

The electron capacities of $Q_a, Q_b, \ldots$ have been explored by illuminating reaction centers in the presence of excess cytochrome and measuring the absorbance changes that reflect oxidation of cytochrome and formation of reduced quinones. If secondary quinones are blocked (by *o*-phenanthroline or at low temperature) or are missing, each reaction center can effect the transfer of just one electron from cytochrome to Q_a. When present and not blocked, Q_b and "pool" UQ can receive two electrons apiece from the pool of excess cytochrome.

Another way to measure electron capacities, of general applicability in photosynthesis research, is to monitor the variations of fluorescence of Bchl or Chl during constant illumination. This is illustrated in Fig. 8.1 for reaction centers illuminated with an excess of an electron donor such as cytochrome. At the start of illumination all the reaction centers are in the state PIQ_a, and the yield of their fluorescence (measured at 900 nm) has the minimal value f_0. Illumination drives electrons from the donor through the reaction center to the quinones. After the pool of secondary quinones (including Q_b) has been filled to capacity, and Q_a has also been reduced, the reaction centers are all in the state PIQ_a^-, and the fluorescence has the maximum value f_m. Between f_0 and f_m, the fluorescence f is a linear measure of the proportion of reaction centers in which Q_a is reduced: $(f - f_0)$ is proportional to $[PIQ_a^-]$, rising from zero at the start of illumination to $f_m - f_0$, when all Q_a is reduced. The quantity $f_m - f$ measures the proportion of "active" reaction centers, in the state PIQ_a. In this example, with no significant transfer of excitation energy between different reaction centers, the photochemical efficiency for quinone reduction (ϕ_p) is proportional to the fraction of reaction centers that are active;[3] $\phi_p \propto [PIQ_a] \propto f_m - f$. Then if the reaction centers are absorbing quanta at a rate IA, where I is the incident flux and A is the fraction ab-

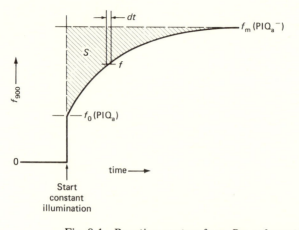

Fig. 8.1. Reaction centers from *Rp. sphaeroides*, mixed with an excess of an electron donor such as cytochrome, are illuminated and the fluorescence at 900 nm is monitored as a function of time. The rise of the fluorescence from f_0 to f_m signals the conversion of the reaction centers from the state PIQ_a to PIQ_a^-. The area S "above" the curve is proportional to the total number of electrons transferred from the donor into Q_a and secondary quinones, in converting all the reaction centers to PIQ_a^-. (See analysis in the text.)

sorbed, the rate of electron transfer from the donor to the quinones is $d(e^-)/dt = IA\phi_p$. If I and A are constant, and $\phi_p \propto f_m - f$, we have $d(e^-)/dt \propto f_m - f$. The total number of electrons transferred to Q_a and secondary quinones, during the conversion of all the reaction centers from PIQ_a to PIQ_a^-, is $\int [d(e^-)/dt] \, dt = \text{const.} \times \int (f_m - f) \, dt$. This is proportional to the area S "above" the fluorescence curve as shown in Fig. 8.1. The constant of proportionality can be evaluated by adding a known amount of quinone to the pool (for example, one UQ per reaction center) and repeating the experiment. If electrons are leaking out of the pool (returning to the oxidized donor) at an appreciable rate, the measurement will be in error. This problem can be overcome by using light that is strong enough to fill the pool much faster than it is being emptied. Experiments of this kind with reaction centers from *Rp. sphaeroides* have shown that the electron capacity of Q_a is 1.0 and that of Q_b (and other secondary UQ) is 2.0, confirming the results of direct observation of cytochrome oxidation as described earlier.

Redox titrations of Q_a have been made by exposing reaction centers or membrane fragments to various ambient potentials and then assaying the state of Q_a. The proportion of oxidized Q_a is indicated by the capacity for light-induced conversion of PIQ_a to $P^+IQ_a^-$, measured most easily by the absorbance changes of $P \longrightarrow P^+$ that persist more than a few milliseconds after a flash.

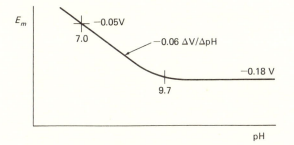

Fig. 8.2. Effect of pH on the redox midpoint potential of the primary quinone Q_a in *Rp. sphaeroides*.

In chromatophores of *Rp. sphaeroides*, the midpoint potential E_m of Q_a is -0.05 V at pH 7 ($n = 1$), varying with pH as shown in Fig. 8.2. The dependence on pH shows that in this kind of equilibrium titration, with Q_a exposed for minutes to the ambient redox potential, there is time for Q_a^- to interact with protons, $Q_a^- + H^+ \rightleftharpoons Q_aH$. The pH of 9.7 represents a pK of the quinone. At higher pH the protonation of Q_a^- is negligible, and E_m has the constant value -0.18 V, representative of Q_a/Q_a^-. When Q_a^- is formed photochemically, it passes its electron to Q_b in a time (<1 msec) too short to allow significant interaction with protons, even at pH 7. Under physiological conditions this quinone shuttles between only the states Q_a and Q_a^-; therefore the operating E_m of Q_a is probably -0.18 V, even though the equilibrium titrations give more positive values.

Isolated reaction centers of *Rp. sphaeroides* show E_m for Q_a/Q_a^- equal to -0.05 V ($n = 1$), independent of pH between 6 and 9, and this poses a problem. The independence of pH indicates that in the isolated reaction centers, H^+ does not have access to Q_a^- even on the time scale of an equilibrium titration. We should then expect $E_m = -0.18$ V, as in the chromatophores at high pH. This problem has not been resolved; it serves as a reminder that reaction centers in isolation do not behave exactly as they do in their natural setting.

The redox properties of Q_a in *Rp. sphaeroides* are fairly representative of most species of purple bacteria, even though Q_a is menaquinone rather than ubiquinone in some species.

With E_m of $P/P^+ = 0.4$ to 0.5 V for these bacteria, and with E_m of $Q_a/Q_a^- = -0.15$ to -0.2 V under operating conditions (no protonation), the transfer of an electron from P to Q_a spans a difference of about 0.6 V in midpoint potential.

The foregoing use of redox potentials is based on the theory of reversible thermodynamics. In this theory each particle (molecule, electron, etc.) is part of an ensemble of many particles, and the behavior of the ensemble is governed by the statistical interactions of the particles. Changes from one state to another can be described as progressions through an infinitude of tiny reversible steps, and the parameters (free energy, entropy, etc.) for the initial and final states are computed by visualizing such progressions through equilibrium states. The early steps of electron transfer in photosynthesis cannot be described properly in these terms. The electron carriers in separate reaction centers do not interact with each other as a statistical ensemble, except perhaps at the level of secondary pools of quinones and cytochromes. The steps are irreversible; that is how the photochemical separation of charge becomes stabilized. The dynamics of electron transfer might in principle be predicted from the theory of irreversible thermodynamics of open systems, but it has been more profitable to take a microscopic point of view, describing the flow of electrons from one molecule to another in terms of quantum states of the molecules (by electron tunneling and nuclear vibrational relaxation). Even so, the redox properties of the electron carriers as found in equilibrium titrations are a useful guide to the rationality of electron transfer sequences and H^+ transport across membranes. We shall return to these considerations of energetics in Part IV.

The ESR spectrum of Q_a^- in *Rp. sphaeroides* was initially a source of confusion. It has peaks at $g = 1.82$ and 1.68, suggesting a bound iron atom and not a quinone. It was then discovered that reaction centers or membrane fragments depleted of iron show an ESR spectrum typical of $UQ^-\cdot$, and that the substitution of Mn for Fe yields a spectrum with still a different form. The deletion of Fe and the substitution of Mn did not appear to alter the functioning of Q_a as an electron acceptor. Furthermore, studies of the Mössbauer effect[4] showed that the iron in reaction centers retains a valence of Fe^{2+}, regardless of whether Q_a is oxidized or reduced. A reasonable conclusion is that the ESR spectrum of Q_a^- in reaction centers is distorted by magnetic (not electronic) interactions with the iron. The function of iron is to facilitate electron transfer from Q_a^- to Q_b. Selective extraction of iron from reaction centers prevents this transfer.[5] The loss of iron does not alter the reactions $PIQ_a \longrightarrow P^+I^-Q_a \longrightarrow P^+IQ_a^-$.

In purple bacteria the immediate donor of electrons to P^+ is a cytochrome of the *c* type. There are usually two or more such cytochromes in close contact with the reaction center, either bound firmly in the membrane or lying next to it in the periplasm. In the latter case, found in *Rp. sphaeroides* and *Rs. rubrum*, breaking the cells allows some of this cytochrome to float away, but some is trapped inside the vesicular chromatophores where it can interact with P^+. This soluble cytochrome, called Cyt c_2, has in *Rp. sphaeroides* an absorption maximum at 551 nm (α band) in the reduced form, and $E_m = +0.30$ V ($n = 1$), independent of pH. In species such as *Rp. viridis* and *Chr. vinosum* the cytochromes that react directly with P^+ are not soluble in water; they are bound in the membranes, and they remain bound to the isolated reaction centers. The bound *c*-type cytochromes of *Chr. vinosum* include two molecules of "high-potential" Cyt 555 (α band maximum 555 nm; $E_m =$

+0.34 V) and two of "low-potential" Cyt 553 (553 nm; +0.1 V). Reaction centers of *Rp. viridis* retain two molecules of Cyt 558 (558 nm; +0.3 V) and three of Cyt 552 (552 nm; 0 V). It has been speculated that the high- and low-potential cytochromes mediate cyclic and noncyclic patterns of electron transfer, respectively.

Various chemical tricks have been used, especially by G. Feher and M. Y. Okamura, to probe the sites of interaction between reaction centers and cytochromes and quinones. In general, molecules including proteins can be allowed to associate with each other as they will, and these associations can then be made permanent by generating covalent bonds between the molecules. Reaction centers of *Rp. sphaeroides* (with component polypeptides H, M, and L; see Section 5.1) have been mixed with cytochrome (either native or mammalian), fixed by cross-linking the proteins, and then subjected to electrophoresis in polyacrylamide. The relative abundances of different couples (L·H, L·M, Cyt·L, etc.) have shown that polypeptide H is closer to L than to M, and cytochrome associates preferentially with L and M and not with H. Turning to the quinones, the "primary" Q_a can be replaced by a quinone analogue that is active as an electron acceptor, is radioactive, and can be linked covalently to the protein by a photochemical process ("photoaffinity labeling"). Electrophoresis of reaction centers treated in this way has shown that the quinone is bound preferentially to polypeptide M.

Antibodies prepared against reaction centers and their subunits can be tested for their reactions with each surface of the membrane.[6] These studies have shown that subunit H is exposed to the outer surface of the chromatophore membrane but not to the inner surface. Some studies have indicated that subunits L and M are "visible" from both sides, but there is diagreement on this.

Mild denaturation of the isolated reaction centers removes subunit H while leaving the Bchl, Bpheo, and Q_a, with photochemical activity intact, associated with the combined subunits L and M. Finally, the directionality of H^+ transport coupled to cyclic electron transport, and also the locus of the Cyt c_2 (inside the chromatophores), shows that P870 is close to the inner surface of the chromatophore membrane, and the binding of protons by reduced quinons happens at the outer surface.

These pieces of information can be assembled in a model for the components of reaction centers in the membranes of *Rp. sphaeroides*, as shown in Fig. 8.3.

Returning to the question of early photochemical events in purple bacteria, the first evidence for an electron carrier between the special pair P and the quinone Q_a was found in reaction centers of *Rp. sphaeroides* poised at low redox potential so as to reduce the quinones. In such reaction centers a laser flash was seen to induce two distinct and consecutive absorbance changes, each with its own characteristic spectrum. The first change appeared with the flash (on a time scale of nanoseconds) and decayed with half-time about

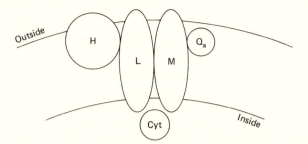

Fig. 8.3. A model for the disposition of the three polypeptides H, L, and M of reaction centers in the intracytoplasmic membranes of *Rp. sphaeroides*, and the sites of the electron acceptor Q_a (ubiquinone) and the cytochrome that donates electrons to oxidized P870. "Outside" refers to the outer surfaces of the chromatophores (intracytoplasmic membrane vesicles). The model is based on experiments described in the text.

10 nsec at room temperature and 20 nsec between 4 K and 150 K. The second change grew in as the first change decayed, and persisted much longer: half-time 6 μsec at room temperature and 110 μsec below 150 K. From their spectra, the first change was attributed to oxidation of the special pair coupled to reduction of Bpheo, and the second to the formation of a triplet state of the special pair (ESR spectra of reaction centers at low redox potential had already shown the light-induced formation of this triplet state). At low temperature both reactions occurred with quantum efficiency near 100%, showing that the first change was a precursor of the second. At room temperature the first step exhibited high quantum efficiency, but the efficiency of triplet formation was only about 15%. These changes fit the following scheme, with superscripts 1 and 3 denoting singlet and triplet states of the special pair:

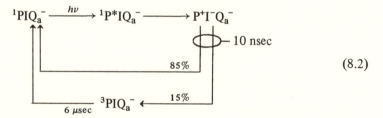

$$(8.2)$$

At low temperature the return from $P^+I^-Q_a^-$ to PIQ_a^- was entirely by way of the triplet state $^3PIQ_a^-$. In subsequent experiments with improved time resolution the state $P^+I^-Q_a$ was detected in reaction centers at physiological redox potential (quinones not reduced in the dark); this state appeared within <10 psec after a flash and decayed to $P^+IQ_a^-$ with half-time 200 psec. These findings, plus the known back-reaction from $P^+IQ_a^-$ to PIQ_a, could be expressed by the following scheme:

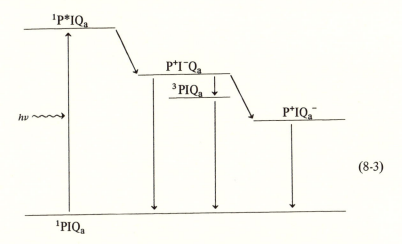

$$(8\text{-}3)$$

The rapid (200 psec) conversion of $P^+I^-Q_a$ to $P^+IQ_a^-$ supervenes over the much slower conversion to $^3PIQ_a$ or to $^1PIQ_a$; so the triplet state is not detected unless Q_a has been reduced chemically and cannot accept an electron from I^-.

These phenomena, with similar kinetic and spectral features, were then found in other species of purple bacteria. A method was also found for trapping the reduced Bpheo (I^-) in a stable form, so that it could be studied at leisure and not as a fleeting transient entity. If reaction centers are illuminated under reducing conditions in the presence of cytochrome as an electron donor, the transfer of an electron from cytochrome to P^+ can compete with the internal processes shown in Reaction 8.2:

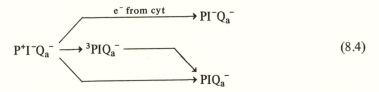

$$(8.4)$$

The competition is strongly against the reaction with cytochrome, which is far slower than the internal reactions. Once it has been formed, however, the state $PI^-Q_a^-$ is stable for many minutes. Thus by repeated cycling of the photochemical system, through thousands of turnovers (requiring a few minutes' exposure to strong light), the reaction centers can be driven predominantly into this state in which I^- has been trapped. The state $PI^-Q_a^-$ reverts slowly to PIQ_a^- as I^- gives up an electron to its surroundings. In *Rp. sphaeroides* the electron can also leave I^- to form the doubly reduced Q_a^{2-},

$PI^-Q_a^- \longrightarrow PIQ_a^{2-}$. The photochemical cycling in the presence of cytochrome must then be continued in order to generate the relatively stable $PI^-Q_a^{2-}$. Evidence for the double reduction of Q_a is less clear in *Rp. viridis* and in *Chr. vinosum*. In these organisms Q_a is menaquinone rather than ubiquinone.

The ESR spectrum of I^- ($g = 2.0025$; band width 12.5 gauss) suggests that the unpaired electron is confined to just one molecule (one Bpheo); this has been confirmed for *Rp. viridis* by ENDOR. In *Chr. vinosum* and *Rp. viridis* the ESR band is split into two at low temperature, probably because the unpaired electron of I^- interacts magnetically with Q_a^-.

The photoconversion of PIQ_a to $P^+I^-Q_a$ is attended by transient absorbance changes near 800 nm, seen during the first 30 psec after a laser flash. This experimental context provides rich ground for optical artifacts,[7] but it is possible (and the idea is appealing) that the voyeur Bchls somehow assist in the transfer of an electron from the special pair to the Bpheo. The wave function of an electron in the excited special pair could spread initially, in vibronically "hot" states, to the voyeur Bchls, and become consolidated mainly in the Bpheo in the course of nuclear relaxation processes (vibrational "cooling"). There are also features near 800 nm in the absorption spectrum of the stable trapped I^-; these features resemble band shifts of Bchl and might be ascribed to the action of local electric fields. Some published spectra of I^- show considerable bleaching near 575 and 800 nm, but this might have been the result of irreversible destruction of Bchl in the harsh procedure for trapping I^-.

Returning to the triplet state that is formed when reaction centers are illuminated at low redox potential, the absorbance change shows that this triplet is confined to the special pair: Formation of the triplet is attended by loss of the 870 nm absorption band, with little change around 800 nm. The ESR spectrum of this triplet is unusual; it shows that the center one of the three magnetically split states is populated much more heavily than the other two. Such "polarization" is not expected when an excited singlet state goes over directly into a triplet; the only rational explanation is that the special-pair triplet is formed from a triplet version of P^+I^-, in which the unpaired spins of P^+ and I^- have lost their original antiparallel (singlet) relationship: $^1P^+I^- \longrightarrow {}^3P^+I^- \longrightarrow {}^3PI$. The triplet $^3P^+I^-$ is expected to be concentrated in the center one of the three triplet levels,[8] and the "memory" of this concentration is carried over into ^3PI.

Taking all of these considerations into account, we can amplify Reaction 8.3 into the scheme shown in Fig. 8.4.

If a carotenoid molecule is close to (or bound to) the reaction center, it will react with the triplet special pair, giving a triplet state of the carotenoid (which decays to the ground state) and returning P to its ground state: $^3P + Car \longrightarrow {}^1P + {}^3Car$.

One would hope that further insight into the reactions shown in Fig. 8.4 can come from studies of the prompt and delayed fluorescence emitted by

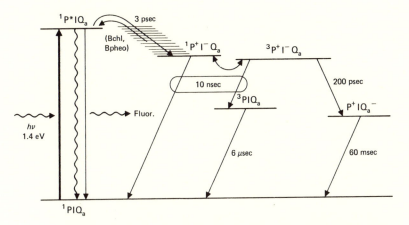

Fig. 8.4. A scheme for photochemical electron transfer in reaction centers of *Rp. sphaeroides*. Parasitic reactions that compete with the formation of $P^+IQ_a^-$ are included, along with approximate half-times for various steps at room temperature. With some modification of these half-times the scheme is applicable to reaction centers from other species of purple photosynthetic bacteria. (Details are discussed in the text.)

reaction centers of purple bacteria. Recall from Section 5.3 that in reaction centers of *Rp. sphaeroides* the yield of fluorescence at 900 nm (corresponding to the absorption band at 870 nm) increases about threefold, rising from 3×10^{-4} to 10^{-3}, when the quinone Q_a is reduced. This might be ascribed to a slowing of the step $P^*I \longrightarrow P^+I^-$ when Q_a is reduced, or to an increased back-reaction, which regenerates the fluorescent state P^*I. The yield of prompt fluorescence should be an indicator of the rate of $P^*I \longrightarrow P^+I^-$, the fluorescence rising if this reaction is slowed. The delayed fluorescence, arising from a back-reaction such as $P^+I^- \longrightarrow P^*I$, should decay with a half-time equal to the half-life of the state (such as P^+I^-) from which the back-reaction originates. In fact the prompt fluorescence increases somewhat when Q_a is reduced, but far less than threefold. The delayed fluorescence, which is appreciable only when Q_a has been reduced or removed from the reaction centers, has a principal component(s) with lifetime in the range 5–10 nsec. However, this delayed fluorescence shows no correlation with the state P^+I^-, or with any known potential precursor in a back-reaction to P^*I. The absence of correlation is shown in studies of temperature dependence and of the effects of an external magnetic field. We cannot yet claim to understand the various manifestations of fluorescence emitted by reaction centers.

The difference in free energy between P^*I and P^+I^- can be estimated from measurements of midpoint redox potentials. If reaction centers of *Rp. viridis* are brought to a potential below about -0.5 V, the state PIQ_a is converted chemically to PI⁻Q$_a$⁻, as evidenced by the failure of light to cause the conversion of I to I⁻. This agrees with a midpoint potential (polarographic half-wave potential) of -0.5 to -0.55 V for Bpheo *b* in vitro. Then with E_m = +0.4 to +0.5 V for P/P⁺ in *Rp. viridis*, the transfer of an electron from P to I spans a difference of about 1 V in midpoint potentials. This can be compared with the energy of P*I, which is 1.24 eV above the ground state if we compute the energy of a quantum at the 1000 nm fluorescence peak. The chemical reduction of I to I⁻ has not been accomplished with reaction centers of bacteria such as *Rp. sphaeroides* or *Chr. vinosum*, which contain Bchl *a* and Bpheo *a*.

The reaction centers of green sulfur bacteria occupy a curious position between those of purple bacteria and those of green plant Photosystem 1. As in the purple bacteria, the primary electron donor appears to be a special pair of Bchl *a* molecules, but operating at a less positive midpoint potential (see Table 8.1). Bpheo *a* has been found in fractions from green bacteria enriched for reaction centers, but the stoichiometry is uncertain. No pheophytin *a* has been found in green plant fractions enriched for Photosystem 1.

The photochemistry of green sulfur bacteria forms a stable reductant of midpoint potential about -0.55 V (independent of pH); this is a strong reductant comparable to that made by Photosystem 1. Unlike the purple bacteria, the green bacteria can effect the direct light-driven reduction of NAD⁺ without resorting to the formation and subsequent utilization of ATP. The reduction of NAD⁺ is reminiscent of NADP⁺ reduction by Photosystem 1 (described in Sections 8.2 and 10.1), in that it is mediated by a soluble ferredoxin.

At least four distinct entities have been detected by ESR in green bacteria, having spectra in their reduced forms like those of the Fe—S—proteins. They have midpoint potentials of +0.16, -0.025, -0.175, and -0.55 V. Not one of them is reduced by light at 4 K, suggesting that none is a "primary" photochemical electron acceptor. The one with E_m = -0.55 V, which is reduced by light at room temperature, could be analogous to the acceptor in Photosystem 1 termed X or A₂ (Section 8.2). The green sulfur bacteria contain a type of menaquinone but no ubiquinone. There is no evidence, however, that quinones participate in the early steps of photochemical electron transfer in these bacteria.

8.2   Reaction centers of Photosystems 1 and 2 of green plants and algae

The presence of antenna Chl has hampered our efforts to study the reaction centers of green plant tissues, but we are beginning to learn the properties of these photochemical centers. For Photosystem 2 there is evidence that Chl *a* (P680) effects the photochemical transfer of an electron to pheophytin *a* (Pheo *a*) and thence to plastoquinone. The similarity to bac-

Table 8.2. *A comparison of photochemical electron donors and acceptors in four types of photosynthetic system*

	Electron donor	Electron acceptors
Photosystem 2	Chl *a* (monomer or special pair) (>+0.8 V)	Pheo *a*, Q_a, Q_b (plastoquinones)
Purple bacteria	Bchl *a* special pair +0.4 V	Bpheo *a*, Q_a, Q_b
Green sulfur bacteria	Bchl *a* special pair +0.3 V	(Bpheo *a*), Fe—S—proteins
Photosystem 1	Chl *a* special pair +0.4 V	(Chl *a*), Fe—S—proteins

terial photosynthesis is striking until we consider that the electron donor P680 must have a midpoint potential (P/P^+) more positive than about +0.8 V if it is to initiate the removal of electrons from water at a pH not too far from 7. The difference could well be that P680 is a monomer of Chl *a* in a nonaqueous environment (see Section 7.2); this could give P/P^+ a midpoint potential up to +0.9 V while preventing a premature reaction of P^+ with water. We know almost nothing about the entities that carry electrons from H_2O and act as donors to oxidized P680 but the process is not a direct reaction between P^+ and H_2O. It is a controlled sequence in which the H^+ and O_2 liberated from H_2O are deposited inside the thylakoids. The O_2 diffuses out.

In Photosystem 1 the photochemical electron donor, P700, is more nearly like the P870 of purple bacteria, having E_m = 0.4 to 0.5 V and using cytochromes to some degree as electron donors. But the primary and secondary electron acceptors are neither Pheo *a* nor quinones; for the most part they are iron-sulfur-proteins that can stabilize the reductant at a potential close to -0.6 V, as in the green sulfur bacteria. Although Bpheo *a* has been found in "reaction center fractions" of green sulfur bacteria, its function has not been established. In Photosystem 1 there is evidence that Chl *a* acts in place of Pheo *a* as a primary electron acceptor.

These comparative aspects of photochemistry, on which we shall enlarge, can be summarized as in Table 8.2, with the more uncertain aspects in parentheses. These similarities and differences provide interesting points of departure for speculations on the evolutionary diversification of photosynthetic life.

Properties of the electron donor in Photosystems 1 and 2 are tabulated in Table 8.3, again with parentheses around the more uncertain elements. The extinction coefficient of the 703 nm band of P700 has been evaluated through a coupled reaction with cytochrome ($Cyt_{red} + P^+ \longrightarrow Cyt_{ox} + P$), as for the bacterial P870. The value is 120 mM^{-1} cm^{-1}.

Photosystem 1 delivers electrons to a soluble ferredoxin, and ultimately

Table 8.3. *Properties of photochemical electron donors in Photosystems 1 and 2*

| | Light-induced absorbance change (peak, nm) | E_m (V) | ESR spectrum | |
			g Value	Band width (gauss)
Photosystem 1, P700	$-680, -703^a$ (+686), +810	+0.4 to +0.5	2.0025	7–8
Photosystem 2, P680	-682 +825	(>+0.8)	2.0026	~8

^{a}697 nm in Photosystem 1 fractions isolated with the help of detergents.

to $NADP^+$ (noncyclic mode) or else through a cycle back to P700. Ferredoxins are Fe—S—proteins that appear in many pathways of electron transport, especially in anaerobic (not necessarily photosynthetic) bacteria. The structures of these Fe—S centers have been elucidated by chemical and physical techniques including Mössbauer spectroscopy. The structures encountered most frequently among various ferredoxins are of the 2Fe,2S and 4Fe,4S types as shown in Fig. 8.5. Each Fe in these structures has four S atoms as ligands. Some bacterial ferredoxins are 8Fe,8S; these have a pair of closely associated 4Fe,4S centers bound to a single protein. The soluble ferredoxin of spinach is 2Fe,2S.

In the oxidized form of a ferredoxin all the iron atoms are ferric. In the reduced form a single electron is acquired by one of the Fe atoms in 2Fe,2S types, and is delocalized over two or more Fe atoms in 4Fe,4S types. Fe—S centers are extremely versatile as electron carriers. The E_m values of various Fe—S—proteins vary widely, between about +0.3 and -0.6 V. The close association of two 4Fe,4S centers on one protein raises the possibility that such ferredoxins can provide a "one electron, two electron" interface between other carriers.

Fe—S centers bound to Photosystem 1 were first detected by light-induced changes of ESR spectra. Various techniques served to reveal ever more "primary" Fe—S centers, and finally an electron acceptor acting before any of these centers. The more peripheral acceptors could be peeled away by selective attack on the reaction center, principally by means of detergents. Their activity could also be deleted by progressive chemical reduction, the more peripheral acceptors being the most easily reduced. In this process the midpoint potentials of the various Fe—S centers were determined. One or more peripheral acceptors could also be reduced photochemically, using an electron donor to restore the oxidized P700 to its reduced form after the photochemistry: $P \cdots A \xrightarrow{h\nu} P^+ \cdots A^-$, followed by $P^+ + \text{donor} \longrightarrow P + \text{oxidized}$

Fe — S Centers

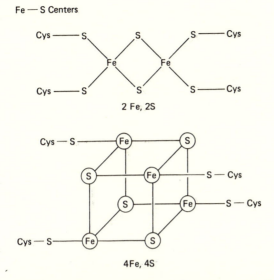

2 Fe, 2S

4Fe, 4S

Fig. 8.5. The structure of Fe—S centers encountered in ferredox-ins, common electron carriers in bacteria and plants. Cys—S refers to a cysteine residue in the protein to which the center is bound. The S atoms that are not part of cysteine are released upon mild acidifi-cation; there are two of these (the central ones) in 2Fe,2S centers and four in 4Fe,4S centers. The smell of H_2S released by acidifica-tion provides a sensitive test for such Fe—S centers.

donor. This quest for earlier acceptors has been aided by improved time reso-lution, in both absorbance and ESR measurements, and by the use of low temperature to cut off the flow of electrons to later acceptors and to assess the temperature dependence (or independence) of the early steps. In some experiments the reduction of an early electron acceptor could not be seen directly, but its presence could be inferred if light caused P700 to become oxidized, and all previously known acceptors had been deleted or blocked.

A current scheme of electron acceptors in Photosystem 1, with arrows showing the movements of electrons, is:

$$P700 \longrightarrow A_1 \longrightarrow A_2 \text{ (or X)} \longrightarrow [A, B] \longrightarrow$$

soluble ferredoxin $\longrightarrow$ to $NADP^+$ and to cycle   (8.5)

X is a synonym for A_2, and the set of acceptors A and B has collectively been called P430 (a term that may have included A_2 in some experiments). X has also been misapplied to A_1. The kinetics of the early steps are not well defined; an electron moves from A_2^- to [A, B] within 20 μsec.

Electrons are donated to P^+ by cytochromes and the copper-containing plastocyanin; this topic will be discussed in Section 9.2.

Table 8.4. *Properties of secondary electron acceptors in Photosystem 1*

Acceptor	ESR bands (g values)	E_m (V)
A_2	1.78, 1.90, 2.07	-0.7
A	1.87, 1.95, 2.05	-0.54
B	1.89, 1.93, 1.96	-0.59

Acceptors A and B show a light-induced decrease of absorbance at 430 nm, signaling their reduction (hence the term P430). They are identified more clearly by the ESR spectra of their reduced forms; these spectra also distinguish them from the earlier acceptor A_2 (or X) (see Table 8.4). All of these ESR spectra are characteristic of Fe—S centers. The redox potentials shown in the table were found by exposing Photosystem 1-enriched fractions to various potentials and assaying the state of reduction of each component by its ability to receive an electron photochemically from P700. A and B are believed to be 4Fe,4S centers. Acidification of Photosystem 1 fractions releases 10-12 sulfur atoms (see the legend of Fig. 8.5), suggesting that A_2 is also a 4Fe,4S center, but possibly a 2Fe,2S center. Close interaction between A and B is indicated by observations of changes in their ESR spectra. If A is already reduced, the subsequent reduction of B not only brings on the ESR spectrum of B, but also modifies the spectrum of A. This has led K. Sauer and collaborators to propose that A and B comprise an 8Fe,8S cluster, two 4Fe,4S centers in close proximity. We have no convincing explanation of the exceptionally low redox potential of A_2. The electron capacities of A_2, A, and B are one each, judging from the amount of P700 (or associated electron donor) that is oxidized when each acceptor is reduced photochemically.

We do not have sufficient information to exclude "parallel branch" modifications of the linear sequence shown in Reaction 8.5. It has been proposed on circumstantial grounds that A and B operate in parallel, with A sending electrons to $NADP^+$ and B mediating cyclic electron transport:

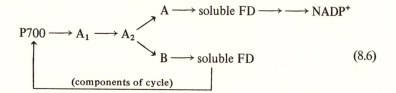

$$(8.6)$$

This would imply that there are two separate pools of soluble ferredoxin. Redox titrations of the soluble ferredoxin give $E_m = -0.38$ to -0.42 V; it is a 2Fe,2S protein.

The distance from A_2 to P700 has been estimated to be more than 1.5 nm because the reduction of A_2 does not perturb the ESR spectrum of P^+. P^+ and A_2^- engage in a back-reaction of half-time 250 μsec; for P^+ reacting with A^- or B^- the half-time is about 30 msec (all at room temperature).

The earliest known acceptor A_1 is not well characterized. It is manifested by the reversible oxidation of P700 under reducing conditions such that A_2, A, B, and all more peripheral acceptors are reduced, and at low temperature (5 K) as well as at room temperature. The signs (ESR and absorbance) of flash-induced P700 oxidation are accompanied by other manifestations attributable to the reduction of A_1: an ESR band with $g = 2.00$ and absorbance changes appropriate (at least crudely) for the reduction of Chl a. It has therefore been suggested that A_1 is Chl a situated close to the special pair, P700. Some preliminary experiments with laser flashes suggest that A_1 receives an electron from P700 within about 10 psec after a flash. A back-reaction between P^+ and A_1^- has half-time 3 μsec at room temperature. The midpoint potential of A_1 is not known; for Chl a in vitro it falls in the range -0.8 to -0.9 V.

In living tissue the flow of electrons through A_1 and A_2 is rapid, and these entities do not accumulate appreciably in their reduced forms. For the most part the reaction center of Photosystem 1 cycles among states in which P700 and peripheral acceptors become oxidized or reduced. States such as $P \cdots A$, $P^+ \cdots A$, $P \cdots A^-$, and $P^+ \cdots A^-$ (or with B and B^- as another variable) all have about the same efficiency in quenching excitation energy, so there is little change in antenna fluorescence during conversion among these states. If Photosystem 1-enriched particles are placed in a reducing environment (as with sodium dithionite), the acceptors A and B become reduced in the dark. Moderate illumination then reduces the earlier acceptor A_2; $PA_1A_2 \longrightarrow P^+A_1A_2^-$. The reducing milieu causes P^+ to become re-reduced, giving the state $PA_1A_2^-$. Fluorescence of the Photosystem 1 subantenna increases about twofold in the change from PA_1A_2 to $PA_1A_2^-$; it has an intermediate value in the state $P^+A_1A_2^-$.

Combining the present knowledge of acceptors in Photosystem 1 with Bengis and Nelson's studies of protein components (Section 6.1) and with the directionality of electron and proton movement through the thylakoid membrane, both Sauer et al. and Bengis and Nelson have proposed similar models for the Photosystem 1 reaction center, as shown in Fig. 8.6.

Turning to Photosystem 2, the reaction center shows a striking similarity to the reaction centers of purple bacteria. As in Reaction 8.1, the main path of electron flow after photo-excitation can be written

$$P680 \longrightarrow I \text{ (Pheo } a) \longrightarrow Q_a \longrightarrow Q_b \longrightarrow \text{Quinone pool} \qquad (8.7)$$

The major points of difference are: (1) P680, a monomer or a special pair of Chl a, has $E_m > +0.54$ V as found in titrations of its ESR signal, and probably $>+0.8$ V if it is to take electrons ultimately from water at a physiologi-

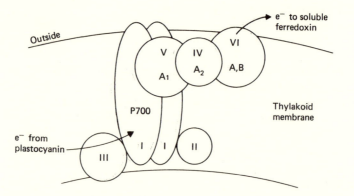

Fig. 8.6. A synthesis of models for the reaction center of Photosystem 1 and its relationship to electron donors and acceptors in the thylakoid membranes of spinach. Protein components I–VI as described by C. Bengis and N. Nelson (C. Bengis and N. Nelson, 1977, *J. Biol. Chem.* 252, 4564–9) have the following weights (kdalton), based on electrophoresis in polyacrylamide: I, 70; II, 25; III, 20; IV, 18; V, 16; VI, 8. In a model proposed by K. Sauer et al. (K. Sauer, P. Mathis, S. Acker, and J. A. van Best, 1978, *Biochim. Biophys. Acta 503*, 120–34) the electron acceptors A_1, A_2, and [A, B] occupy the positions shown by components V, IV, and VI, respectively, in the model of Bengis and Nelson. The electron donor P700 is associated with two copies of Component I. Plastocyanin will donate electrons to oxidized P700, but only if Component III is present. The role of Component II is not known.

cal pH. (2) The component I is Pheo a; the evidence for its position in Reaction 8.7 remains circumstantial (see next paragraph). (3) The quinones are all plastoquinones (PQ); this is certified by the spectra of light-induced absorbance changes in which Q_a or Q_b is reduced to the anionic semiquinone (PQ⁻·), and by extraction and chemical characterization of the quinones. (4) The donors to oxidized P680 are for the most part unidentified; the pattern of their reactions with P^+ remains confusing.

Evidence for the participation of Pheo a as the electron carrier I is incomplete; we lack pertinent measurements of light-induced absorbance changes on a picosecond time scale. Pheo a is found in Fraction 2 preparations of chloroplasts. The state $PI^-Q_a^-$ has been generated in such preparations, as in the reaction centers of purple bacteria, by prolonged illumination at low redox potential (−0.05 to −0.5 V) in the presence of an electron donor (see Reaction 8.4). The presumed conversion from PIQ_a^- to $PI^-Q_a^-$ is signaled by a spectrum of absorbance changes suggesting the reduction of Pheo a: a decrease at 545 nm, a very broad increase in the red, and appropriate changes in the blue (Soret band region). There are sharp negative bands at 676 and 685 nm; the possible involvement of voyeur Chl a has not been established.

The fluorescence of Chl a associated with the reaction centers of Photosystem 2 has been a major tool in investigating these reactions. Numerous correlations with light- and chemically induced absorbance changes have established that the states $P680 \cdot Q_a$, $P680^+ \cdot Q_a$, and $P680^+ \cdot Q_a^-$ are all strongly quenching states, giving a low yield of fluorescence. Only the state $P680 \cdot Q_a^-$ gives high fluorescence.[9]

The reactions involving P and Q_a can be analyzed by a combination of absorbance and fluorescence measurements. In the back-reaction from $P^+ Q_a^-$ to PQ_a, the absorbance changes reflect $P^+ \longrightarrow P$ and $Q_a^- \longrightarrow Q_a$, and the fluorescence does not change because $P^+ Q_a^-$ and PQ_a are both low fluorescence states. If P^+ is reduced to P by an electron donor other than Q_a^-, the state $P^+ Q_a^-$ is converted to PQ_a^-, and the fluorescence rises. Then if Q_a^- passes its electron to Q_b, while P remains reduced ($PQ_a^- Q_b \longrightarrow PQ_a Q_b^-$), the fluorescence falls again. These techniques have revealed the following kinetics (reaction half-times) at room temperature, with arrows again showing the movements of electrons:

$$Z \xrightarrow{\text{30 nsec}} P^+ \underset{?}{\xleftarrow{\text{130–200 } \mu sec}} Q_a^- \xrightarrow{\text{200–600 } \mu sec} Q_b \qquad (8.8)$$

The donor Z (alias D and Y in the scientific literature) is the most rapid donor to P^+ known; it is on the path of electron transfer from water (see Chapter 9).

We do not know how rapidly an electron moves from the presumed I^- to Q_a, or back to P^+; this determination must await picosecond absorption measurements pertaining to I/I^-.

The reduction of Q_a is shown most directly by an absorbance decrease at 260 nm and an increase at 320 nm, typical of plastoquinone reduction ($PQ \longrightarrow PQ^- \cdot$). In analogy to the purple bacteria, this reaction is attended by other absorbance changes resembling band shifts of Pheo a, both in the red (near 680 nm) and around 545 nm. The latter effect, an increase at 543 nm and a decrease at 547 nm, has been termed C550.

If chloroplasts are extracted with hexane containing a little methanol, so as to remove the plastoquinones and carotenes, the signs of photochemical reduction of Q_a (including C550) are lost. If PQ is added back, the activity (and presumably the presence) of Q_a is restored, but C550 remains lost. Adding β-carotene restores C550. This does *not* mean that C550 represents a band shift of the carotene; the positive and negative bands at 543 and 547 nm are too sharp to correspond to β-carotene. The carotene appears to play a part in allowing the reduction of Q_a to influence the spectrum of Pheo a. As in the purple bacteria (Fig. 5.11c), this "electrochromic" reaction of Pheo a has nothing to do with its possible role as a photochemical electron acceptor.

The transfer of an electron from Q_a^- to Q_b is blocked by 3-(3,4-dichlorophenyl)-1,1-dimethylurea (DCMU), an action similar to that of o-phenanthroline in purple bacteria. DCMU can also shift electrons back from Q_b^- to Q_a^-. A deficiency of bicarbonate ion also slows electron transfer from Q_a^- to Q_b; in

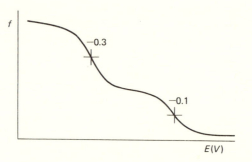

Fig. 8.7. Fluorescence yield of spinach chloroplasts as a function of ambient redox potential.

earlier times this effect was misinterpreted to signify an involvement of CO_2 in "the light reaction" (Reaction 1.2).

The simple picture of a linear path of electron flow from water to the PQ pool, passing through Z, P, I, Q_a, and Q_b, has been confounded by several experiments suggesting that other electron donors and acceptors can act in parallel with Z and Q_a respectively:

$$(8.9)$$

At present we understand neither the chemical identities nor the physiological significance of these alternate donors and acceptors.

The earliest evidence that something might be wrong with the simple linear sequence came from measurements of the fluorescence of chloroplasts exposed to different redox potentials, in an effort to conduct an equilibrium redox titration of Q_a/Q_a^-. High fluorescence could be expected for the state PQ_a^- and low fluorescence for PQ_a. A plot of fluorescence (the component that varies with redox potential) against the ambient redox potential showed two waves centered at about -0.1 and -0.3 V at pH 7 (Fig. 8.7). Both inflections, showing midpoints at -0.1 and -0.3 V, varied with pH according to $\Delta E_m/\Delta pH = -0.06$, implying the binding of H^+ on the time scale of an equilibrium titration. From this experiment there appeared to be two distinct molecules (Q_a and Q'), or else two redox states of Q_a (singly and doubly reduced), affecting the fluorescence in an additive way. Subchloroplast fractions enriched for Photosystem 2 show only the more negative component. We are not certain which component in this equilibrium titration should be identified with Q_a/Q_a^- in chloroplasts. The other component might not be operative on the time scale of electron transport under physiological conditions.

More decisive evidence for an alternate acceptor Q' was found with chloroplasts that had been exposed to a high concentration (1 M) of tris(hydroxymethyl) aminoethane (Tris, a commonly used pH buffer). This treatment interrupts the flow of electrons from water to the reaction center, but electrons can be delivered to P^+ from an external donor such as phenylene diamine, $H_2N-Ph-NH_2$. At intense flash of light lasting about 2 μsec (too short to allow Q_a^- to be formed *and* to discharge an electron to Q_b) causes the transfer of two electrons from phenylene diamine, through P680, and on to the reducing side. One electron goes to Q_a and on to Q_b. The other electron presumably goes to an alternate acceptor, Q'. A second flash, given after Q_a^- has had time to pass its electron to Q_b but before the hypothetical $(Q')^-$ has released its electron, causes the transfer of just one electron, from phenylene diamine to Q_a.

In a complicated assortment of experiments made with chloroplasts that have not been denatured, alternate donors and acceptors have also been implicated (see P. Joliot and A. Joliot in the Suggested Readings). The experiments have involved observations of fluorescence; of absorbance changes for P/P^+, Q_a/Q_a^-, and the "C550" band shift; and of oxygen evolution in response to sequential flashes and to continuous illumination. They show that an alternate acceptor, perhaps the same Q' that has been inferred from the experiments mentioned earlier, can become reduced under some conditions. The redox state of this acceptor affects the fluorescence, in the same way as the redox state of Q_a but less strongly. The C550 effect occurs only when Q_a is reduced. Q' appears to be a cul-de-sac; there is no evidence that it can transfer electrons rapidly into the mainstream of electron flow from Photosystem 2 to Photosystem 1. At "natural" light intensities the arrival of excitation quanta at a Photosystem 2 reaction center is no more frequent than once every few milliseconds, and Q_a^- has ample time to pass its electron to Q_b before another quantum comes along. The experiments with Tris-washed chloroplasts indicate that Q' remains reduced for minutes after it has taken an electron. Therefore the normal acceptor is probably Q_a, with Q' participating only in the rare event that excitation reaches a reaction center in the state $(P)_{Q_a^-}^{Q'}$.

The reduction of the alternate acceptor Q' is coupled to the oxidation of an alternate donor D, another cul-de-sac in the sense that its oxidation does not promote oxygen evolution. Electron donation from Z to P^+ has a half-time of 30 nsec; that from D to P^+ goes with half-time about 20 μsec. It may be that every reaction center has two donors and two acceptors, $_Z^D(P)_{Q_a}^{Q'}$, or that some of the reaction centers have DPQ' and others have ZPQ_a. In the latter case the two kinds of reaction center compete for the same pool of quanta in the antenna, and ZPQ_a is the stronger competitor.

Hydroxylamine is an especially potent inhibitor of electron transport from water to P^+. In its presence the half-time for reduction of P^+ becomes 20 μsec, regardless of the state of the oxygen-evolving system. It is as if the reaction center and its donors have become fixed in the "slow" state. Hydroxylamine removes manganese from chloroplasts; one possibility is that Z *is* Mn^{2+}, and its removal forces the sole use of the alternate donor D. Externally added Mn^{2+} is an effective donor to P^+ in Tris-denatured chloroplasts. The problem is complicated by the fact that hydroxylamine itself is an excellent donor to P^+, possibly with half-time 20 μsec.

For many years a cytochrome of the *b* type, Cyt 559, was presumed to be an electron carrier between Photosystems 1 and 2. There is no sound evidence for this. A light-induced oxidation of Cyt 559 can be observed, coupled to the re-reduction of $P680^+$, but only at low temperature or in chloroplasts that have been deranged, as by exposure to Tris buffer. At present we do not see a convincing function for Cyt 559.

These problems involving multiple donors and acceptors in Photosystem 2 are far from being resolved. It is intriguing to imagine that the donor D and the acceptor Q' are components of an electron transport cycle that includes Cyt 559.

SUGGESTED READINGS

Clayton, R. K. (1978). Physicochemical mechanisms in reaction centers of photosynthetic bacteria. In *The Photosynthetic Bacteria*, R. K. Clayton and W. R. Sistrom, eds., Chapter 20, pp. 387–96. Plenum, New York.

Parson, W. W., Clayton, R. K., and Cogdell, R. J. (1975). Excited states of photosynthetic reaction centers at low redox potentials. *Biochim. Biophys. Acta 387*, 265–78.

Fajer, J., Davis, M. S., Brune, D. C., Spauling, L. D., Borg, D. C., and Forman, A. (1977). Chlorophyll radicals and primary events. *Brookhaven Symp. Biol. 28*, 74–103.

Feher, G., Hoff, A. J., Isaacson, R. A., and Ackerson, L. C. (1975). ENDOR experiments on chlorophyll and bacteriochlorophyll in vitro and in the photosynthetic unit. *Ann. N.Y. Acad. Sci. 244*, 239–59.

Clayton, R. K., and Straley, S. C. (1972). Photochemical electron transport in photosynthetic reaction centers. IV. Observations related to the reduced photoproducts. *Biophys. J. 12*, 1221–34.

Parson, W. W. (1969). The reaction between primary and secondary electron acceptors in bacterial photosynthesis. *Biochim. Biophys. Acta 189*, 384–96.

Clayton, R. K., Fleming, H., and Szuts, E. Z. (1972). Photochemical electron transport in photosynthetic reaction centers from *Rhodopseudomonas sphaeroides:* II. Interaction with external electron donors and acceptors and a reevaluation of some spectroscopic data. *Biophys. J. 12*, 46–63.

Feher, G., and Okamura, M. Y. (1977). Reaction centers from *Rhodopseudomonas sphaeroides. Brookhaven Symp. Biol. 28*, 183–93. (1978). Chemical composition and properties of reaction centers. In *The Photosynthetic Bacteria*, R. K. Clayton and W. R. Sistrom, eds., Chapter 19, pp. 349–86. Plenum, New York.

Prince, R. C., and Dutton, P. L. (1976). The primary acceptor of bacterial photosynthesis; its operating mid-point potential? *Arch. Biochem. Biophys. 172*, 329–34.

Dutton, P. L., Leigh, J. S., Jr., and Reed, D. W. (1973). Primary events in the photosynthetic reaction centre from *Rhodopseudomonas sphaeroides* strain R26: Triplet and oxidized states of bacteriochlorophyll and the identification of the primary electron acceptor. *Biochim. Biophys. Acta 292*, 654–64.

Blankenship, R. E., and Parson, W. W. (1979). The involvement of iron and ubiquinone in electron transfer reactions mediated by reaction centers from photosynthetic bacteria. *Biochim. Biophys. Acta 545*, 429–44.

Loach, P. A., and Hall, R. L. (1972). The question of the primary electron acceptor in bacterial photosynthesis. *Proc. Natl. Acad. Sci. U.S. 69*, 786–90.

Marinetti, T., Okamura, M. Y., and Feher, G. (1979). Photoaffinity labeling of the quinone binding site in bacterial reaction centers of *Rhodopseudomonas sphaeroides*. *Biophys. J. 25*, 204a.

Leigh, J. S., and Dutton, P. L. (1974). Reaction center bacteriochlorophyll triplet states: Redox potential dependence and kinetics. *Biochim. Biophys. Acta 357*, 67–77.

Shuvalov, V. A., and Klimov, V. V. (1976). The primary photoreactions in the complex cytochrome-P890-P760 (bacteriopheophytin$_{760}$) of *Chromatium minutissimum* at low redox potentials. *Biochim. Biophys. Acta 440*, 587–99.

Tiede, D. M., Prince, R. C., and Dutton, P. L. (1976). EPR and optical spectroscopic properties of the electron carrier intermediate between the reaction center bacteriochlorophylls and the primary acceptor in *Chromatium vinosum*. *Biochim. Biophys. Acta 449*, 447–67.

Jortner, J. (1976). Temperature-dependent activation energy for electron transfer between biological molecules. *J. Chem. Phys. 64*, 4860–7.

Akhmanov, S. A., Borisov, A. Yu., Danielius, R. V., Kozlovskij, V. S., Piskarskas, A. S., and Razjivin, A. P. (1978). Primary photosynthesis selectively excited by tunable picosecond parametric oscillator. In *Picosecond Phenomena* (Proc. First International Conference of Picosecond Phenomena, Hilton Head, South Carolina, U.S.A.), C. V. Shank, E. P. Ippen, and S. L. Shapiro, eds., pp. 134–9. Springer-Verlag, New York.

Thurnauer, M. C., Katz, J. J., and Norris, J. R. (1975). The triplet state in bacterial photosynthesis: Possible mechanisms of the primary photo-act. *Proc. Natl. Acad. Sci. U.S. 72*, 3270–4.

Katz, J. J., Shipman, L. L., and Norris, J. R. (1979). Structure and function in photoreaction-centre chlorophyll. In *Chlorophyll Organization and Energy Transfer in Photosynthesis* (Ciba Foundation Symposium 61), pp. 1–34. Excerpta Medica (Elsevier), Amsterdam.

Clarke, R. H., Connors, R. E., and Frank, H. A. (1976). Investigation of the structure of the reaction center in photosynthetic bacteria by optical detection of triplet state magnetic resonance. *Biochem. Biophys. Res. Commun. 71*, 671–5.

Rubin, A. B. (1978). Picosecond fluorescence and energy transfer in primary photosynthetic processes. *Photochem. Photobiol. 28*, 1021–8.

Campillo, A. J., and Shapiro, S. L. (1978). Picosecond fluorescence studies of excitation migration and annihilation in photosynthetic systems. *Photochem. Photobiol. 28*, 975–89.

Knaff, D. B. (1978). Reducing potentials and the pathway of NAD⁺ reduction. In *The Photosynthetic Bacteria*, R. K. Clayton and W. R. Sistrom, eds., Chapter 32, pp. 629–40. Plenum, New York.

Olson, J. M., Prince, R. C., and Brune, D. C. (1977). Reaction-center complexes from green bacteria. *Brookhaven Symp. Biol. 28*, 238–45.

Sauer, K. (1979). Photosynthesis—The light reactions. *Ann. Rev. Phys. Chem. 30*, 155–78.

Bearden, A. J., and Malkin, R. (1975). Primary photochemical reactions in chloroplast photosynthesis. *Q. Rev. Biophys. 7*, 131–77.

Kok, B. (1961). Partial purification and determination of oxidation reduction potential of the photosynthetic chlorophyll complex absorbing at 700 mμ. *Biochim. Biophys. Acta 48*, 527–33.

Malkin, R., and Bearden, A. J. (1975). Laser-flash-activated electron paramagnetic resonance studies of primary photochemical reactions in chloroplasts. *Biochim. Biophys. Acta 396*, 250–9.

Malkin, R., and Bearden, A. J. (1971). Primary reactions of photosynthesis: Photoreduction of a bound chloroplast ferredoxin at low temperature as detected by EPR spectroscopy. *Proc. Natl. Acad. Sci. U.S. 68*, 16–19.

Sauer, K., Mathis, P., Acker, S., and van Best, J. A. (1978). Electron acceptors associated with P-700 in Triton solubilized Photosystem 1 particles from spinach chloroplasts. *Biochim. Biophys. Acta 503*, 120–34.

Shuvalov, V. A., Dolan, E., and Ke, B. (1979). Spectral and kinetic evidence for two early electron acceptors in Photosystem 1. *Proc. Natl. Acad. Sci. U.S. 76*, 770–3.

Telfer, A., Barber, J., Heathcote, P., and Evans, M. C. W. (1978). Variable chlorophyll *a* fluorescence from P-700 enriched Photosystem 1 particles dependent on the redox state of the reaction centre. *Biochim. Biophys. Acta 504*, 153–64.

Klimov, V. V., Klevanik, A. V., Shuvalov, V. A., and Krasnovsky, A. A. (1977). Reduction of pheophytin in the primary light reaction of Photosystem 2. *FEBS Lett. 82*, 183–6.

Amesz, J., and Duysens, L. N. M. (1977). Primary and associated reactions of System II. In *Primary Processes of Photosynthesis*, J. Barber, ed., Chapter 4, pp. 151–85. Elsevier/North Holland, Amsterdam.

Butler, W. L. (1973). Primary photochemistry of Photosystem II of photosynthesis. *Acc. Chem. Res. 6*, 177–84.

Joliot, P., and Joliot, A. (1979). Comparative study of the fluorescence yield and of the C550 absorption change at room temperature. *Biochim. Biophys. Acta 546*, 93–105.

9 Oxygen evolution; secondary transport of electrons and protons

9.1 Photosynthetic oxygen evolution

In the reaction centers of Photosystem 2, one quantum can effect the transfer of just one electron from the donor P680 to an acceptor. The oxidizing equivalents produced by four such photoacts must cooperate somehow in promoting the reaction

$$2H_2O \xrightarrow{\ 4\ h\nu\ } 4e^- + 4H^+ + O_2 \qquad (9.1)$$

The electrons released in this reaction flow to P680$^+$. In earlier times this four-quantum cooperation could be formulated sensibly in a variety of ways; for example, a pair of "two-quantum" reactions $H_2O \xrightarrow{\ 2\ h\nu\ } 2e^- + 2H^+ + O$ followed by $O + O \longrightarrow O_2$. These formulations could require different degrees of interaction between the oxidizing entities made by separate reaction centers. We shall see, however, that there is no such interaction; each Photosystem 2 reaction center must generate four oxidizing equivalents in four consecutive photoacts, and these oxidizing equivalents must be stored and then used by the chemical machinery associated uniquely with that reaction center.

The first evidence that oxygen evolution requires the summation of more than one photoact came in two forms. First, the rate of O_2 evolution by algae in continuous weak light is not a linear function of the light intensity. Measurements suggested a quadratic (rate $\propto I^2$) or higher order of dependence; the most refined measurements indicated a fourth-power dependence in weak light. This is consistent with a sequential four-quantum mechanism, for example:

$$S_0 \underset{\ }{\overset{h\nu}{\rightleftharpoons}} S_1 \underset{\ }{\overset{h\nu}{\rightleftharpoons}} S_2 \underset{\ }{\overset{h\nu}{\rightleftharpoons}} S_3 \underset{\ }{\overset{h\nu}{\rightleftharpoons}} S_4$$
$$\downarrow$$
$$O_2 \qquad (9.2)$$

for which the rate (R) of O_2 evolution is expected to vary with light intensity (I) according to $R = I^4/(a + bI + cI^2 + dI^3 + eI^4)$, which reduces to $R \propto I^4$ when I is small.

More dramatic evidence for the summation of one-quantum events came from studies that measured the changing yields of oxygen evolved after consecutive flashes of light applied to dark-adapted algae. To make such measurements it was necessary to measure the small amount of O_2 evolved by a sample (suspension of algae) of manageable size in response to a single brief flash, an amount of the order of 10^{-10} mole. This was first achieved by Allen and Franck,[1] who used the quenching of the phosphorescence of acridine adsorbed on silica as a sensitive assay for O_2. Allen and Franck found that a dark-adapted suspension of *Chlorella* produced no O_2 in response to a single intense submillisecond flash. A second flash given a few seconds after the first yielded some O_2, and subsequent flashes, spaced a few seconds apart, gave "normal" yields of O_2. These experiments demonstrated an activation requirement: a light-dependent step (or steps) that prepared the algae for subsequent light-driven oxygen evolution. The activated state was lost after a few minutes of darkness.

The next significance advance was made in the 1960s by P. Joliot, using a conventional polarographic method (an oxygen electrode), but of improved design. The use of modulated light, and hence a modulated response of the electrode, afforded better discrimination of signal from noise, and the geometry and materials of the electrode minimized the time needed for O_2 to diffuse from the algae (again *Chlorella*) to the electrode. Joliot could measure the rate of O_2 evolution in very weak continuous (but modulated) light, and later the yield of O_2 from single flashes. The initial rate of O_2 evolution in weak continuous light could be measured after prolonged darkness or after one or more intense "conditioning" flashes; this gave essentially the same knowledge as a measurement of the flash yield on consecutive flashes.

First Joliot confirmed the "activation" requirement and showed, from the dependence on wavelength or by the use of inhibitors, that this effect of light is mediated solely by Photosystem 2. Photochemical activity of Photosystem 1 plays no part in priming the algae for O_2 evolution. Joliot then showed, using flashes of a few microseconds' duration, that very little O_2 is evolved in response to the first two flashes after dark adaptation. Thus two conditioning acts are needed before the algae are primed and ready to release O_2 when illuminated further. Allen and Franck had observed some O_2 evolution in response to the second flash (i.e., after only one conditioning flash) because their flashes were long enough to produce a significant proportion of "double hits": two turnovers of a reaction center during one flash.

The most striking result of Joliot's experiments was that the O_2 evolved on consecutive flashes showed an oscillating pattern with a periodicity of four: when consecutive flashes were given to dark-adapted *Chlorella*, the yield of O_2 was maximal in response to flashes number 3, 7, 11, and so on.[2] This pattern, sketched in Fig. 9.1 for a similar experiment made with spinach chloroplasts, showed damping such that eventually every flash gave the same yield. Note that the first two flashes gave almost no O_2.

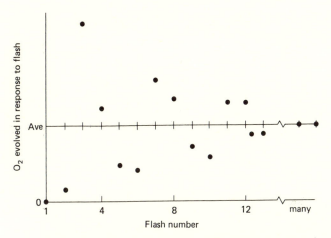

Fig. 9.1. Oxygen evolution by spinach chloroplasts, in response to consecutive saturating flashes of light after 40 min dark-adaptation. Flashes were 2 sec, spaced 0.3 sec apart. The horizontal line labeled "Ave " shows the yield after many flashes. Adapted from B. Forbush, B. Kok, and M. McGloin (1971), *Photochem. Photobiol. 14*, 307–21.

To explain this behavior, B. Kok (with whom Joliot had collaborated) proposed a linear four-quantum mechanism, a variant of Reaction 9.2 that has come to be called the "cycle of S-states" (Fig. 9.2). States S_0, S_1, and so on are consecutively higher oxidation states of a chemical system; a photoact by Photosystem 2 removes an electron from S_0 and converts it to S_1, and so forth. State S_4 is unstable; it releases O_2 and reverts spontaneously to S_0. In a complete cycle, requiring four photoacts, two molecules of H_2O are taken up, four electrons are delivered to the reaction center of Photosystem 2, four H^+ are released (inside the thylakoid; see later in this section) and one O_2 is liberated. To fit his and Joliot's data, Kok proposed that S_3 and S_2 decay slowly to S_1. These back-reactions account for the loss of "activation" during dark adaptation. The system is primed or activated when it is in the state S_3; a single photoact can then produce oxygen: $S_3 \xrightarrow{h\nu} S_4 \longrightarrow S_0 + O_2$.

The damping of the oscillating pattern shown in Fig. 9.1 is explained by a small proportion of double hits and misses in the reaction centers; double hits advance the phase of the cycle, and misses retard it. The result is that after many flashes, or prolonged illumination, the population of S-systems has become randomized so that 25% are in each of the states S_0, S_1, S_2, and S_3. After a few minutes' dark-adaption, S_3 and S_2 have reverted to S_1; the distribution is then 25% S_0, 75% S_1, and no S_2 or S_3. After two flashes the 75% in state S_1 have been promoted to S_3 and are ready to release O_2 in response to another flash. These changing distributions are summarized (in idealized

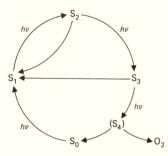

Fig. 9.2. The cycle of S-states. (See text for discussion.)

Table 9.1. *Cycle of S-states: predicted distributions (percent in each state)*
during a basic experimental protocol[a]

Experimental condition	S_0	S_1	S_2	S_3
Randomized (after prolonged illumination)	25	25	25	25
Dark-adapted after having been randomized	25	75	0	0
After first flash following dark adaptation	0	25	75	0
After second flash	0	0	25	75
After third flash	75	0	0	25
After fourth flash	25	75	0	0

[a]Effects of double hits and misses have been ignored.

form, ignoring double hits and misses) in Table 9.1. The percent in each state
is listed, and the percent in state S_3 shows how much O_2 will be released by
the next flash.

Kok could account accurately for the data shown in Fig. 9.1 by invoking
5% double hits (due to the finite duration of each flash) and 10% misses (due
to reaction centers that were temporarily inactive, perhaps in the state PQ_a^-
or $P^+Q_a^-$). In other experiments, using 30-nsec flashes rather than 2-μsec
flashes, the data could be fit without double hits, with only misses that retard
the progression through the cycle.

The S-cycle is a plausible and economical hypothesis that has satisfied a
great variety of experimental tests, most of them variations of the basic
experiment of Fig. 9.1. One example is provided by the protocol of sequential operations given in Table 9.2. By the trick of giving one flash between
two periods of dark adaptation, the population of S_1 is made 100%. Two subsequent flashes put every system into the state S_3. As expected, a third flash
then gives the maximum possible yield of O_2, more than that from the third
flash in Fig. 9.1.

Each S-system is associated uniquely with one reaction center of Photosystem 2; it cannot give electrons to other reaction centers. The oxidizing equiv-

Table 9.2. *Cycle of S-states: predicted distributions during another experimental protocol*[a]

	Predicted distribution among S-states after each operation			
Operation	S_0	S_1	S_2	S_3
Illumintae several minutes	25	25	25	25
Dark-adapt	25	75	0	0
Give one flash	0	25	75	0
Dark-adapt again	0	100	0	0
Give one flash	0	0	100	0
Give second flash	0	0	0	100
Give third flash	100	0	0	0

[a] Effects of double hits and misses have been ignored.

alents made in a single flash by two or more reaction centers cannot combine to promote one S-system through more than one step. This absence of cooperativity between reaction centers has been shown with chloroplasts treated with DCMU and illuminated so as to render most of the Photosystem 2 reaction centers inactive, fixed in the state PQ_a^-. With only 3% of the reaction centers active, so that the potential for cooperation between reaction centers is greatly diminished, the yield of each flash is decreased in the same proportion; the flash yield pattern has the same form as when nearly all of the reaction centers are active.

Analyses of the flash yield pattern and of the flow of electrons through the reaction centers to plastoquinone have shown that each of the four steps ($S_0 \longrightarrow S_1$, etc.) has the same quantum efficiency, with one electron transferred in each step. The kinetics of the S-cycle have been explored by varying the period of dark adaptation and the interval between flashes, and noting the effects of these variations on the yield of O_2 from each flash. If consecutive flashes are spaced too closely together, the flashes become less effective in promoting the S-states. Originally this observation was formulated by separating each step, such as $S_0 \longrightarrow S_1$, into a "light" and a "dark" phase; $S_0 \xrightarrow{h\nu} S_0' \xrightarrow{dark} S_1$. A flash could promote S_0 to S_0', but the promotion of S_1 by a subsequent flash had to await the completion of $S_0' \longrightarrow S_1$. The progress of each of these dark steps could thus be monitored by measuring the effectiveness of a flash given a known short time after a preceding one. With suitable protocols each of the four steps in the S-cycle could be isolated. The half-times of $S_0' \longrightarrow S_1$, $S_1' \longrightarrow S_2$, and $S_2' \longrightarrow S_3$ all proved to be in the range 200–500 μsec. The kinetics of these hypothetical reactions paralleled the conversion of PQ_a^- to PQ_a in the reaction centers, as inferred from the decline of the yield of fluorescence (see Section 8.2 and Reaction 8.8). In view of this correlation, the "dark" step should be identified with recovery of the reaction center, and not with a reaction in the S-cycle. Instead of writing $S_0 \xrightarrow{h\nu} S_0' \longrightarrow S_1$, it would seem more sensible to write

(1) $PQ_a \longrightarrow P^+Q_a^-$

(2) $P^+Q_a^- + S_0 \longrightarrow PQ_a^- + S_1$ (9.3)

(3) $PQ_a^- \longrightarrow PQ_a$

The second of these reactions, in which an electron is moved from S_0 to P^+, is much faster than the third, which must be completed before another flash can be effective. The conversions of S_1 to S_2 and S_2 to S_3 can be formulated similarly. The sequence of reactions $S_3 \xrightarrow{h\nu} S_4 \longrightarrow S_0 + O_2$ is in a separate category; about 1 msec is needed for the completion of these steps. We can speculate that the release of O_2 is a rate-limiting event, slower than the recovery of PQ_a^- to PQ_a. This is consistent with Joliot's observations of oxygen evolution by algae exposed to modulated light. From the difference in phase between the modulated light and the modulated appearance of O_2, he concluded that there is a rate-limiting step of about 1 msec half-time between the photochemistry and the final release of O_2.

Deactivation, the spontaneous return of S_3 and S_2 to S_1, has been studied by giving varying periods of dark-adaptation, preceded and followed by sequences of flashes. In chloroplasts the kinetics suggest that S_3 decays to S_2, and S_2 to S_1, with half-times about 1 min each. In *Chlorella* the half-times are about 10 sec for $S_3 \longrightarrow S_2$ and 30 sec for $S_2 \longrightarrow S_1$. These reactions are reductions, and the sources of electrons for these reductions are not clear. External electron donors can accelerate the conversions of S_3 and S_2 to S_1.

If these deactivation reactions are allowed to proceed from an initially random distribution of S-states (after prolonged illumination), the distribution approaches 25% S_0 and 75% S_1 within a few minutes. Other regimes of dark-adaptation interspersed with flashes of light can give other distributions, such as 100% S_1, as we have seen. But after several hours the distribution approaches a final equilibrium of 20% S_0 and 80% S_1, regardless of the distribution reached earlier in consequence of the decay of S_3 and S_2 to S_1. Evidently there is a very slow interconversion of S_0 and S_1, and the final ratio of 80 S_1:20 S_0 reflects a redox equilibrium between these states. Addition of a mild reductant, reduced indophenol dye, shifts the final equilibrium to 50 S_1:50 S_0, and the oxidant ferricyanide yields a final state of nearly 100% S_1, with very little S_0.

We have used the cycle of S-states as a formalism for describing the storage of four oxidizing equivalents when four electrons are driven sequentially through a reaction center of Photosystem 2. We should not restrict our thinking by visualizing the S-system as a single molecular complex in which all of the chemical events of Reaction 9.1 occur. One possibility, consistent with current knowledge, is that an electron carrier Z connects a water-binding and O_2-releasing system (S) to the reaction center:

$$S \longrightarrow Z \longrightarrow P680 \xrightarrow{h\nu} Q_a \text{ (plastoquinone)} \tag{9.4}$$

In this picture the storage of positive charge and the release of H^+ could involve Z as well as S. We shall consider what is known about these components, and about the partial reactions (movements of electrons and release of H^+) that occur in each of the four steps of the cycle.

When P^+ has been formed photochemically ($PQ_a \xrightarrow{h\nu} P^+Q_a^-$), it is re-reduced by the hypothetical donor Z ($ZP^+ \longrightarrow Z^+P$) with half-time 30 nsec. This process has been monitored by the decay of an absorption band at 825 nm (characteristic of P^+) as well as by the rise in fluorescence that attends the conversion of $P^+Q_a^-$ to PQ_a^- (Section 8.2; Reaction 8.8). No direct manifestation of Z, or of $Z \longleftrightarrow Z^+$, has yet been discovered.

A hint about the flow of electrons from S to Z^+ exists in the form of an ESR signal ($g = 2.00$) called Signal II_{vf}. The spectrum of this signal is not distinctive enough for a chemical identification. The light-induced formation of Signal II_{vf} in chloroplasts after various chemical treatments shows that it reflects the oxidation of some entity by Photosystem 2, and the kinetics show that it is not a manifestation of Z^+. Following a brief flash, Signal II_{vf} is formed with half-time about 20 μsec, whereas P^+ reverts to P (and Z is converted to Z^+) in only 30 nsec. A component with rise time 20 μsec may have a place in the scheme of Reaction 9.4; there is circumstantial evidence that in some phases of the S-cycle the re-reduction of P^+ requires about 20 μsec rather than 30 nsec. We shall return to this point later.

The release of H^+ that attends oxygen evolution has been measured by recording changes of pH, both inside and outside the thylakoids, in response to consecutive flashes of light. The external pH can be measured with a conventional glass electrode or with an indicator such as phenol red, which does not permeate the thylakoids. A few pH indicators such as neutral red apparently penetrate to the interior of the thylakoids and can respond to changes of internal pH. By measuring changes of absorbance at selected wavelengths, one can monitor the responses of these indicators with excellent time resolution.

In these studies with chloroplasts the release of H^+ that attends O_2 evolution must be discriminated from the binding and release of H^+ attending the flow of electrons through plastoquinone, from Photosystem 2 to Photosystem 1. In the latter process, which will be described in the next section, protons are bound from outside the thylakoid when PQ is reduced, and released inside when the PQ is reoxidized. The net result is a transfer of H^+ from outside to inside the thylakoids, except for those protons that are detained in PQH_2. The evolution of O_2 is accompanied by a net release of H^+ inside the thylakoids. The elucidation of these patterns of proton binding, release, and translocation has been aided by the use of chemicals (uncouplers; see Chapter 10) that render the thylakoid membrane freely permeable to H^+. In the presence of such an uncoupler the protons released in consequence of O_2 evolution diffuse to the outside, and any active translocation of protons across the membrane is nullified by passive diffusion.

The release of H^+ inside the thylakoids also changes the intensity of delayed fluorescence, arising from Photosystem 2 through a reversal of the photochemical charge separation. The knowledge gained by studying this delayed fluorescence can be coordinated with the more direct indications of changing H^+ concentration.

Application of these techniques to chloroplasts exposed to consecutive flashes of light has shown that protons are released inside the thylakoids in the pattern

$$S_0 \xrightarrow{} S_1 \longrightarrow S_2 \xrightarrow{} S_3 \xrightarrow{(S_4)} S_0$$
$$\quad\searrow \qquad\qquad\qquad \searrow \quad \searrow$$
$$\quad H^+ \qquad\qquad\qquad H^+ \quad 2H^+$$

$$(9.5)$$

An electron is removed, through the reaction center, in each step of the cycle. Therefore the system (possibly including Z) undergoes no net change in its electric charge in going from S_0 to S_1 (loss of $1e^-$ and $1H^+$). The positive charge increases by one in going from S_1 to S_2 (loss of $1e^-$ only), does not change between S_2 and S_3, and decreases by one (loss of $1e^-$ and $2H^+$) in reverting from S_3 through S_4 to S_0.

There is indirect evidence, based on the change of fluorescence that reflects $P^+Q_a^- \longrightarrow PQ_a^-$, that the reduction of P^+ is fast (30 nsec) by electrons released in $S_0 \longrightarrow S_1$ and $S_1 \longrightarrow S_2$, and slow (about 20 μsec) in the steps $S_2 \longrightarrow S_3$ and $S_3 \longrightarrow S_4 \longrightarrow S_0$. This evidence is tenuous and has not been supported by direct observation of $P^+ \longrightarrow P$ (by the change of absorbance at 825 nm) in the context of consecutive flashes designed to advance the S-cycle.

Combining all of these facts, hints, and speculations, one can make various outlines of the chemical cycle of O_2 evolution. Figure 9.3 is an example of such modeling. In this scheme the light reaction $P \longrightarrow P^+$ is of course accompanied by $Q_a \longrightarrow Q_a^-$ with Q_a^- then delivering its electron to the PQ pool. Steps labeled FAST are of 30 nsec half-time; SLOW means about 20 μsec. It would be premature to lay any claim for the uniqueness or correctness of this model; it is offered only as an example of the sort of speculation that can be made at this time.

The redox potential of the O_2 electrode (Reaction 9.1) is about +0.8 V at pH 7. It follows that the four electrons involved in the reaction must function at an average potential near +0.8 V, and that the couple P680/P680$^+$ should have $E_m \gtrsim$ +0.8 V. In chloroplasts that have been treated so as to abolish the flow of electrons from water (as by washing with Tris buffer; see next para-

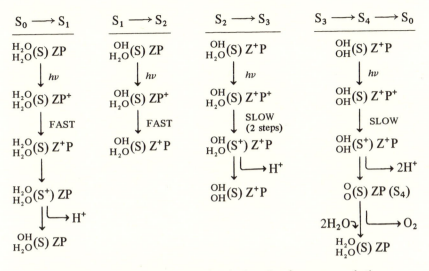

Fig. 9.3. A scheme for the chemical cycle of oxygen evolution.

graph), it can be seen that P^+ is indeed a powerful oxidant. With no electrons coming from water, and with no external electron donor to neutralize P^+, illumination of these chloroplasts causes the indiscriminate oxidation of many components including Chl and carotenoids, presumably through reactions with P^+. Comparable damage is not caused when the donors of Photosystem 1 or purple bacteria are allowed to accumulate in the oxidized form P^+.

Manganese is essential for photosynthetic oxygen evolution, and there is evidence that both S and Z contain this element. The extraction of Mn from chloroplasts yields six atoms for each Photosystem 2 reaction center. Four of these atoms are bound loosely; they can be extracted by washing chloroplasts with Tris buffer. The capacity for O_2 evolution is lost in proportion to the fraction of these four Mn atoms extracted (Fig. 9.4). These atoms may be associated with "S." When all four labile Mn atoms have been extracted, leaving only two that are more firmly bound, O_2 evolution has been abolished. T. Yamashita and W. L. Butler[3] showed that these depleted chloroplasts retain active reaction centers, and electrons can be delivered to P^+ by externally added donors[4] including Mn^{2+}. Hydroxylamine, NH_2OH, removes the two remaining Mn atoms from chloroplasts. It prevents added Mn^{2+} from functioning as a donor to P^+, and it slows the rate of re-reduction of P^+ from 30 nsec to 20 μsec, showing that it interferes with Z. We can speculate that the two firmly bound atoms of Mn comprise an essential part of Z. In addition to Mn, oxygen evolution is dependent on Cl^- or any other monovalent anion of a strong acid.

An advance that holds considerable promise for our understanding of

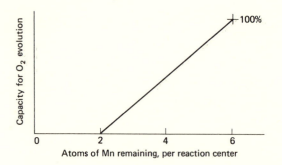

Fig. 9.4. Plot of capacity for oxygen evolution against atoms of manganese remaining, per Photosystem 2 reaction center, under progressive extraction of Mn from spinach chloroplasts.

the chemistry of oxygen evolution has been made by M. Spector and G. D. Winget.[5] Pursuing the technique of fractionating chloroplasts and attempting to recombine the separate components so as to restore biological activity, these investigators used artificial phospholipid vesicles as carriers of the recombined fractions. The vesicles (*liposomes*) are prepared by ultrasonic treatments of soybean phospholipids in aqueous media; they are spherical bodies consisting of a lipid bilayer surrounding an interior aqueous phase. If the liposomes are mixed with proteins and sonicated again, the proteins (especially hydrophobic ones) can be incorporated into the lipid bilayer. The result, a membrane of lipid and protein surrounding an aqueous phase, bears a general resemblance to such biological structures as thylakoids and chromatophores.

Spector and Winget first exposed spinach chloroplasts to 0.1 M sodium cholate (a detergent) and 0.4 M ammonium sulfate; this treatment detached some protein from the thylakoids and abolished oxygen evolution. Centrifugation yielded a supernatant fraction and a dense green pellet of "depleted thylakoids." The pellet was incorporated into liposomes by sonication, and the resulting green liposomes were tested for photochemical activity. They showed no oxygen evolution, but they did perform light-induced electron transfer through both Photosystems 1 and 2, with artificial electron donors and acceptors. Returning to the supernatant from the original centrifugation, Spector and Winget isolated a specific protein weighing 65 kdalton and bearing two Mn atoms. This protein, when incorporated into the green liposomes by sonication, restored photosynthetic oxygen evolution at rates (per unit of chlorophyll) approaching those of the original chloroplasts. The Mn-containing protein in solution associates as dimers when illuminated. In all probability this protein is a major part of the natural O_2-evolving system; perhaps the dimer *is* the system denoted S.

For the first time, then, we appear to be ready to explore the chemistry of the cycle of S-states in detail. This advance came about through the use of liposomes as vehicles for reconstitution, and this technique offers promise in many ways. For example, the separation and recombination of antenna pigment–proteins and reaction centers is probably best attempted with liposomes so as to restore some semblance of the interactions that exist in vivo.

Clearly there is much to be learned about the chemistry of oxygen evolution. The most crucial step in the evolution of plants was the development of an oxidizing system so powerful that water could be used as an abundant supply of electrons in photosynthesis. Yet it is in this area of photosynthesis research that we know the least, and have the greatest room for improving our knowledge.

9.2   Patterns of electron and proton transport surrounding Photosystems 1 and 2

We shall now deal with the pathways of electron and H^+ transport initiated by photochemistry in the reaction centers of Photosystems 1 and 2, as outlined in Figs. 9.5 and 9.6. In Fig. 9.5 the ordinate shows the approximate E_m of each component involved in electron transfer. The flow of electrons is exergonic in all but the light-driven steps P680 $\longrightarrow$ I and P700 $\longrightarrow$ A_1. Electrons will flow as dictated by a tendency toward redox equilibrium (involving protons as well as electrons), constrained by the relative positions of the carriers in the membrane. Figure 9.6 shows roughly how this pattern is arranged in the thylakoid membrane, and where protons are bound, released, and carried across the membrane. Some details of H^+ transport in relation to ATP formation, and of the assimilation of CO_2, will be the subject of Part IV.

This network of reactions has been elucidated mainly by observing light-induced changes of optical absorbance and ESR, while manipulating the material so as to favor one reaction or another. These studies have been supplemented by chemical analyses and observations of Chl fluorescence. The following manipulations, made singly and in various combinations, have proved useful: (1) the use of brief, intense ("single turnover") flashes of actinic light; (2) selection of actinic wavelength to drive Photosystem 1 or Photosystem 2 preferentially; (3) fractionation of chloroplasts and algae into parts enriched for Photosystem 1 or Photosystem 2 activity; (4) adjustment of the ambient redox potential so that components are poised mainly in their oxidized or reduced forms; (5) deletion of selected components by the use of (hopefully) specific inhibitors, by genetic mutation, and by extraction [soluble ferredoxin (FD) and plastocyanin (PCy) can be removed by extraction of broken chloroplasts with aqueous solutions; the quinones can be extracted with hydrocarbon solvents (with or without a trace of methanol)

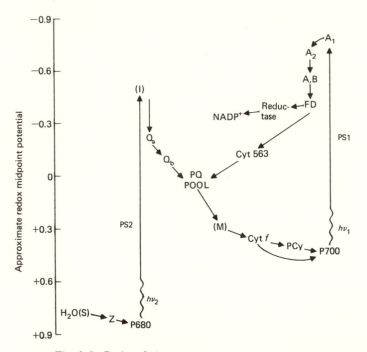

Fig. 9.5. Paths of electron transport in green-plant photosynthesis. The ordinate shows the approximate redox midpoint potential of each component. Components of the sequences H_2O—S—Z—P680—Q_a—Q_b—PQ and P700—A_1—A_2—A,B—FD have been described in Sections 8.2 and 9.1. Parentheses around I and M indicate the tentative nature of these assignments. I is probably pheophytin a; M may be an Fe—S center (see the text). PCy, plastocyanin, is a copper–protein compound. Cyt 563 is a cytochrome of the b type.

without denaturing the rest of the system] ; (6) replacement of deleted components, or substitution with analogous compounds, or addition of artificial electron donors and acceptors.

Most of the research on photosynthesis in the past two decades has been of this nature.

We have considered the evidence (Section 8.2) that pheophytin a, designated I, acts as the earliest photochemical electron acceptor in Photosystem 2. The one-electron chemistries of P680 and I are linked to the "2e⁻, 2H⁺" chemistry of PQ through the special quinones Q_a and Q_b. These special quinones are in an environment such that the anion radical, $Q^-\cdot$, is relatively stable. Absorbance changes attending their reduction indicate that they are both PQ. Extraction of freeze-dried chloroplasts with hexane removes the PQ pool and probably Q_b as well; with the help of a little methanol the activities

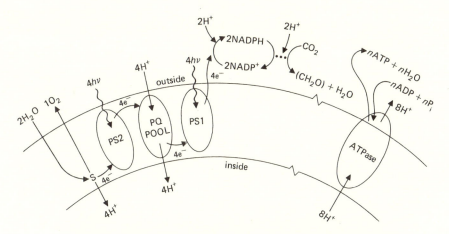

Fig. 9.6. Patterns of electron and H^+ transport in relation to the thylakoid membrane. It is probable but not certain that one H^+ is carried across the membrane for each electron passing through the PQ pool. Four quanta absorbed by each photosystem can then deposit $8H^+$ inside the thylakoid. This is not quite enough to make $3ATP$; it is believed that the ATPase ("coupling factor") converts one ADP to ATP for every $3H^+$, so that n in this figure equals 8/3. P_i denotes H_3PO_4 and (CH_2O) denotes carbohydrate. The cyclic flow of electrons through Cyt 563 and PQ, driven by Photosystem 1 and illustrated in Fig. 9.1, is not shown in this figure. Such cyclic flow through PQ is expected to move H^+ from outside to inside the thylakoid.

associated with Q_a are deleted. Normal activity is restored by adding back PQ. Restoration of the "C550" absorbance change, an indirect indicator of the reduction of Q_a (Section 8.2), requires that β-carotene be added back along with PQ.

Electron transfer from Q_a to Q_b and into the PQ pool is blocked by DCMU and slowed by a deficiency of HCO_3^-. The action of DCMU seems analogous to that of o-phenanthroline in purple bacteria (Section 8.1).

The existence of Q_b as an entity distinct from the rest of the PQ pool, and the pattern of electron flow from Q_a to the pool, have been indicated by the responses of chloroplasts to a sequence of brief flashes of light. Absorbance changes that reflect the formation and disappearance of $PQ^-\cdot$ show an oscillating pattern in chloroplasts exposed to sequential flashes: Following a period of dark adaptation, the semiquinone is formed on the first flash, disappears in consequence of the second flash, reappears on the third, and so on. This pattern of binary oscillations becomes damped after many flashes. We shall go into this kind of behavior in greater detail, drawing on a greater wealth of data, when we consider a similar pattern for the quinones associated with bacterial reaction centers. For now it will suffice to say that the reactions of PQ in chloroplasts support the following scheme, in which PQ represents the plastoquinone pool:

First flash:

$$Q_a Q_b PQ \xrightarrow{h\nu} Q_a^- Q_b PQ \longrightarrow Q_a Q_b^- PQ$$

Second flash:

$$Q_a Q_b^- PQ \xrightarrow{h\nu} Q_a^- Q_b^- PQ \longrightarrow Q_a Q_b^{2-} PQ \xrightarrow{+2H^+} Q_a Q_b PQH_2$$

(9.6)

It is implied that in each flash the light-driven step $Q_a \longrightarrow Q_a^-$ is accompanied by $P \longrightarrow P^+$, followed by the rapid reduction (by Z) of P^+ back to P. In this scheme the primary quinone shuttles only between the states Q_a and Q_a^-. After the first flash, Q_a^- delivers its electron to Q_b in 200–500 μsec. In this "dark" step there is no further change of quinone absorbance because one semiquinone (Q_a^-) disappears while the other (Q_b^-) is formed. The yield of Chl fluorescence declines during this step, concomitant with the return of Q_a^- to Q_a. Once formed, Q_b^- is stable for seconds or longer in the dark. A second flash, given to the system in the state $Q_a Q_b^-$, drives another electron from P680 to Q_a and on to Q_b, forming the double-reduced Q_b^{2-} as evidenced by the disappearance of the semiquinone $(Q^- \cdot)$ absorbance at 320 nm. When Q_b^{2-} passes its electrons to PQ in the pool, $2H^+$ are bound from outside the thylakoid, to form PQH_2. What may happen (see Section 9.3) is that Q_b^{2-} becomes protonated to $Q_b H_2$, falls off into the pool, and is replaced by a new molecule of PQ (from the pool), which assumes the role of Q_b. This restores the system to its original state $Q_a Q_b$, except that one molecule in the pool has been reduced to PQH_2. A third flash then acts like the first flash, starting another binary sequence.

The sequence of Reactions 9.6 predicts that one molecule of semiquinone per reaction center should be formed on flashes 1, 3, 5, . . . and destroyed (converted to Q_b^{2-} and then to PQH_2) on flashes 2, 4, 6, ... The absorbance changes at 320 nm should reflect this. We also expect that no protons are bound after the odd-numbered flashes, and $2H^+$ per reaction center are bound after the even-numbered flashes. Actually the absorbance changes measured in chloroplasts are smaller than expected, and some H^+ is bound after every flash, although the binding of H^+ is greater after even-numbered flashes (Fig. 9.7). These observations satisfy Reactions 9.6 if we assume that in the chloroplasts used for these experiments, after dark adaptation, 70% of the systems were in the state $Q_a Q_b$, and 30% were in the state $Q_a Q_b^-$. The 30% that were "out of phase" would have shown a "second flash" response (Reactions 9.6) on the first flash, nullifying part of the 320 nm absorbance change and giving some H^+ uptake on odd-numbered flashes.

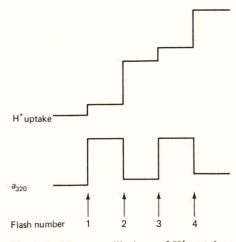

H^+ uptake

a_{320}

Flash number   1   2   3   4

Fig. 9.7. Binary oscillations of H^+ uptake and absorbance at 320 nm, the latter reflecting formation and disappearance of semiquinone, in chloroplasts exposed to consecutive flashes of light.

The first hints of this "$1e^-/2e^-$" interface between the reaction center and the quinone pool were gleaned from measurements with chloroplasts, but the documentation is far more thorough with reaction centers and chromatophores of purple bacteria, as we shall see in the next section.

The size of the PQ pool (molecules per reaction center) has been estimated in several ways: (1) By analyzing the rise in Chl fluorescence during constant illumination (Fig. 8.1), one can deduce how many electrons must be delivered to the pool before Q_b, and finally Q_a, is driven to its reduced form. (2) The pool can be filled by continuous illumination and then "dumped" into an excess of an external electron acceptor such as dichlorophenolindophenol (DCPIP);[6] the amount of DCPIP reduced by the pool is determined. (3) One can extract all of the PQ and determine the ratio of PQ to reaction centers. The first two of these methods require strong illumination, which assures that the filling of the pool by Photosystem 2 activity is much more rapid than its emptying through Photosystem 1. The consensus from all of these approaches is that the pool contains 5-10 molecules, each capable of storing two electrons.

Another question is that of articulation: Does each Photosystem 2 reaction center have its private pool of 5-10 quinone molecules, or do many reaction centers of Photosystem 2 feed electrons into a larger common pool? A tentative answer is given by the fact that if 90% of the reaction centers are blocked from the pool by DCMU, the remaining 10% can reduce nearly all the PQ in the chloroplast. This would indicate that many reaction centers are connected to a common pool, unless the DCMU hops from one site to another so that each reaction center has an open path to its own PQ pool 10% of the time.

The type of articulation between the PQ pool and Photosystem 1, and the nature of a $1e^-/2e^-$ interface between PQ and Photosystem 1, remain subjects of speculation.

Going on from the PQ pool to P700, three distinct electron carriers have been implicated in this pathway: M (hypothetical), cytochrome f, and plasto-cyanin (PCy). M might be an Fe—S center, Cyt f is a cytochrome similar to those of the c type, and PCy is a copper-protein compound. In some algae, notably *Euglena*, Cyt f and PCy are replaced by one or more c-type cyto-chromes. We shall consider the properties of these electron carriers in more detail after we have examined the evidence for their positions in Fig. 9.5.

The positions of M, Cyt f, and PCy in the network between PQ and P700, and indeed the existence of M, were inferred through studies by R. P. Levine[7] on a series of mutants of the green alga *Chlamydomonas*. These mutants were defective in one or another of the electron carriers, and the consequences of these defects could be assessed in suspensions of chloroplasts of the algae. In the normal algae, far red light (delivered almost entirely to the reaction centers of Photosystem 1) first causes the oxidation of P700; the oxidized P700 in turn oxidizes the carriers between itself and Photosystem 2. The components closest to P700, such as PCy and Cyt f (actually Cyt 553, a minor variant of Cyt f in *Chlamydomonas*), are oxidized the most rapidly by this Photosystem 1 activity. Levine found that in mutants lacking PCy, Cyt f was not oxidized by Photosystem 1 (i.e., by P700$^+$). Other mutants, lacking Cyt f, did show a flow of electrons from PCy to P700$^+$, but the normal flow of electrons from Photosystem 2 to PCy did not occur.[8] The properties of these two mutants implied a sequence Photosystem 2 $\cdots \longrightarrow$ Cyt $f \longrightarrow$ PCy $\longrightarrow$ P700. In a third type of mutant electrons could flow to P700 from both PCy and Cyt f, but neither of these carriers could receive electrons from Photosystem 2. Levine concluded that this type of mutant lacks an electron carrier M that stands between Photosystem 2 and Cyt f. Independent evidence for the existence and nature of M will be considered later.

Other attempts to delineate the paths of electron flow from PQ to P700 have included the use of Hg^{2+} as a specific inhibitor of PCy, and kinetic analyses of the oxidation and reduction of Cyt f and PCy following flashes of light that generate P700$^+$. In spinach chloroplasts the kinetic studies have revealed a pattern more elaborate than the linear path proposed by Levine. The rise and decay of the oxidized forms of P700, Cyt f, and PCy, after a flash, show how electrons are transferred among these components and be-tween them and M or PQ. The results remain indecisive; reports from differ-ent laboratories have been somewhat contradictory. Half-times ranging from about 10 to 200 μsec have been reported for the oxidations of both Cyt f and PCy. One interpretation is given in Fig. 9.5; Cyt f may work through PCy or may bypass PCy and feed electrons directly to P700. To be completely conservative one might allow for a network such as

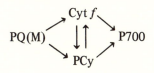

at least in spinach. But in some species of algae the simple linear sequence PQ(M) ⟶ Cyt *f* ⟶ PCy ⟶ P700 appears sufficient.

These differences in the organization of electron transport might be related to differences in the plastocyanins of spinach and green algae. PCy from spinach is a protein made up of four polypeptides (about 10 kdalton each) containing four -SH groups and four atoms of copper. PCy from green algae contains only two Cu atoms per molecule. Each molecule of PCy has an electron-storing and -releasing capacity greater than one; perhaps two in the algae and four in spinach. Thus the PCy may act as a small electron pool between PQ and Photosystem 1, especially if several molecules of PCy are connected jointly to several P700 on one side and to several donor paths (from PQ) on the other. This could account for some of the confusion about the patterns of electron flow between PQ and P700. The oxidation and reduction of PCy is signaled by the reversible bleaching of a broad absorption band centered at 597 nm; also the oxidized form has an ESR band at $g = 2.05$.

Cyt *f* has a narrow absorption band (the α band) at 554 nm in the reduced form; this band disappears when the cytochrome is oxidized. The redox potentials (E_m) of both Cyt *f* and PCy are in the range +0.34 to +0.38 V; that of P700 is about +0.4 V.

R. Malkin and P. J. Aparicio,[9] studying duckweed chloroplasts, have recently described a light-dependent ESR signal that could be a manifestation of the hypothetical M. The spectrum of this signal shows it to be an Fe—S center of a type already known and implicated as an electron carrier in membranes of photosynthetic bacteria, called a Rieske center. The responses of this entity to illumination, and its midpoint potential of +0.29 V, are appropriate for a position between PQ and Cyt *f* as shown in Fig. 9.5.

We do not know how the two-electron chemistry of PQ is connected to the one-electron chemistries of Cyt *f* and P700. When PQH_2 is reoxidized to PQ, $2H^+$ are released inside the thylakoid and two electrons are sent to Photosystem 1. Perhaps these two electrons follow separate paths to two reaction centers (two P700s), or perhaps M can store two electrons and pass them on one at a time. The multi-electron capacity of PCy makes it an attractive candidate for a $2e^-/1e^-$ interface, but direct contact between PQ and PCy has not been demonstrated.

Electrons delivered from P700 to A_1, in the photochemistry of Photosystem 1, make their way rapidly through A_2, A, and B to soluble FD, which might comprise yet another pool, storing electrons at a potential of about -0.4 V. The reduction of $NADP^+$ by reduced FD is mediated by a component, FD-NADP reductase, that could provide an interface between the one-electron chemistry of FD and the two-electron chemistry of NADP. The reaction $NADP^+ + 2e^- + H^+ \longrightarrow NADPH$ occurs on the outside of the thylakoid. The E_m of NADP is -0.35 V at pH 7.

Ferredoxin can also deliver electrons to PQ by way of Cyt 563 (also known as Cyt b_6), completing a cycle of electron flow around Photosystem 1. In this cyclic flow through PQ, the reduction and reoxidation of PQ are attended by the transport of H^+ across the membrane, as in the noncyclic flow from Photosystem 2 through PQ to Photosystem 1. Protons are bound from the outside of the thylakoid, and released on the inside. This proton translocation can provide for ATP formation as a result of the cyclic operation of Photosystem 1.

Arnon and Chain,[10] in the course of presenting new evidence that the cyclic mode plays a significant part in the life of plants, have suggested a mechanism that regulates the partitioning of electron flow between the cyclic and noncyclic modes. When the supply of CO_2 (the ultimate acceptor of electrons from NADPH) is low, the reduced NADPH accumulates. With no $NADP^+$ to receive electrons from FD, these electrons are diverted into the cyclic path, to Cyt 653 and on to PQ. This plausible mechanism can be compared with the speculative one illustrated in Reaction 8.6 (Section 8.2), in which Fe—S centers A and B communicate with separate FD pools, one feeding electrons to $NADP^+$ and the other to the cycle. In this scheme, when CO_2 is low and NADPH builds up, center A ($E_m = -0.54$ V) becomes reduced sooner than center B ($E_m = -0.58$ V). In its reduced state, A cannot receive electrons from A_2. The preferred path of electrons is then through B and into the cycle.

The conclusion that PQ is the acceptor of electrons from Cyt 563 is based primarily on the fact that the cyclic flow of electrons around Photosystem 1 supports ATP formation. This implies that H^+ is transported to the inside of the thylakoid during the cycle, and the only obvious site of such proton translocation is the PQ pool (Fig. 9.6). There are two molecules of Cyt 563 for every Photosystem 1 reaction center; these two acting in concert could provide a $1e^-/2e^-$ interface between FD and PQ. The E_m of Cyt 563 is in the range 0 to -0.18 V; the more negative values are measured in slightly damaged (uncoupled) chloroplasts. The PQ pool operates at a potential near zero.

The foregoing conclusions, embodied in Fig. 9.5, are probably correct for the most part, but many of the details are based on inconclusive evidence and must remain tentative. For example, Cyt 559 was once thought to act as a carrier between Photosystem 2 and Cyt f, but we must now confess ignorance as to its location and function. The position of Cyt 563 as a carrier of electrons from FD to PQ is consonant with observations of the light-induced reduction and oxidation of this cytochrome. However, the kinetic reactions are not easy to correlate with those of the presumed rea ners. New knowledge will probably dictate further revisions of our conception of electron flow around Photosystems 1 and 2.

The reducing side of Photosystem 1 can deliver electrons to substances other than $NADP^+$ and Cyt 563. When $NADP^+$ is in short supply, both O_2 and H^+ can act as electron acceptors. O_2 can be converted, although slowly, to the superoxide radical $O_2^-\cdot$. H^+ can receive electrons from reduced FD; $2H^+ \longrightarrow H_2$. This conversion of H^+ to H_2, using electrons taken from water, is of great interest to those who would exploit solar energy to make fuels. The reduction of H^+ by FD is mediated by the enzyme hydrogenase. This enzyme is rendered inactive by O_2, which poses a technical problem in the development of this type of solar energy conversion.

The same hydrogenase can allow H_2 to act as an electron donor to Photosystem 1, supporting the reduction of $NADP^+$ and ultimately the assimilation of CO_2:

$$\left.\begin{array}{c} H_2 \\ \\ H^+ \end{array}\right\} \longrightarrow (\text{Photosystem 1}) \longrightarrow \left(\begin{array}{c} NADPH \\ \\ NADP^+ \end{array}\right.$$

This reaction, discovered by H. Gaffron, was of major importance in the acceptance of van Niel's views of the chemical nature of photosynthesis (recall Section 1.4).

A great variety of chemicals can act as artificial electron donors or acceptors, interacting with various components of the pathways shown in Fig. 9.5. If a chemical is to interact with a natural component as an electron donor or acceptor, it must meet two requirements, one of energy and one of access. First, the transaction must not be endergonic. A successful donor must have E_m more negative than that of its target, and a successful acceptor must have an E_m more positive. Second, the external reagent must be able to penetrate lipids, if necessary, to reach components inside the thylakoid membrane.

The positions of electron carriers in the membrane, indicated in Fig. 9.6 (also see Fig. 8.6), have been inferred mainly from the loci of H^+ binding and release, from the access of specific antibodies to components on one side or the other of the membrane, and from the presence of identifiable components that are external to the thylakoid, or can be washed off its surface. Further information comes from the distribution of electric charge in the membrane when components are reduced or oxidized (ways to assess this distribution will be described in Chapter 10). Finally, some topological knowledge can be gained from the actions of external electron donors and acceptors. In general, hydrophilic molecules interact well with the reducing side of Photosystem 1, whereas a molecule must be somewhat lipophilic in order to react well with components between the two photosystems and on the oxidizing side of Photosystem 2. The quinones Q_a and Q_b are believed to lie close to the outer surface of the thylakoid membrane, but a lipoidal barrier seems to shelter them from the external aqueous environment. The PQ pool evidently spans the membrane, as it carries H^+ across it, and yet it interacts only slowly with hydrophilic electron donors and acceptors. For example, ferricyanide, $Fe(CN)_6^{3-}$, is a far better acceptor from FD than from any component between Q_a and P700. Acceptors and donors that react well with the chain between the two photosystems include:

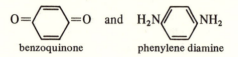

benzoquinone   phenylene diamine

The most effective known donor to the oxidizing side of Photosystem 2, from S to P680, is hydroxylamine, NH_2OH. Other good donors in this region are

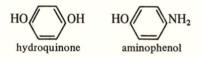

hydroquinone   aminophenol

and phenylene diamine. Certain electron acceptors have E_m values so negative that they can take electrons only from the reducing side of Photosystem 1. Methyl viologen, for example, is reduced at a potential about -0.6 V.

Phenazinium methyl sulfate (PMS), of E_m near zero, is unusual in that in its reduced form it can donate electrons directly to $P700^+$, and in its oxidized form it can accept from FD and perhaps from (A, B). PMS effectively short-circuits the physiological cycle, transferring electrons directly from (A, B) or FD to P700. Surprisingly, the operation of this artificial cycle around Photosystem 1 supports ATP formation. Apparently the PMS acts like PQ in that it binds H^+ from outside the thylakoid when it is reduced and releases H^+ inside when it is reoxidized. Other lipophilic electron carriers such as DCPIP can participate in artificial cycles of electron flow around Photosystem 1, and can translocate H^+ in the process, but they cannot deliver electrons directly to P^+: They must donate their electrons through PCy. The direction of H^+ transport, from outside to inside, is presumably assured not by any special property of the carrier but by the fact that P700 is near the inside of the membrane, and (A, B) and FD are near or on the outside. The carrier must merely be able to diffuse through the membrane.

If we take Fig. 9.6 at face value, eight quanta (four in each photosystem) generate two molecules of NADPH and deposit eight protons inside the thylakoid. A present consensus (Chapter 10), subject to change, is that three H^+ must be moved out of the thylakoid (through the ATPase) in order to convert one ADP to ATP. The assimilation of one molecule of CO_2 by the reductive pentose cycle (Chapter 11) requires two NADPH and three ATP.

The fruits of eight quanta, two NADPH and 8/3 ATP, are not quite enough. But by means of the cyclic path a single quantum in Photosystem 1 can translocate one more H^+, raising the yield of H^+ to nine and providing for three ATP.

Actually the number of protons translocated when electrons flow through the PQ pool remains controversial. One H^+ per electron, as depicted for the PQ pool in Fig. 9.6, is the generally accepted value, and reports of $2H^+/e^-$ have been criticized on methodological grounds. However, if the latter stoichiometry should prove to be correct, then eight quanta in the noncyclic mode would give $12H^+$ rather than 8 inside the thylakoid. This would provide for four ATP, one more than the requisite three, and would make the cyclic mode superfluous for CO_2 assimilation. Nevertheless, plants need ATP for many energy-requiring activities beyond the conversion of CO_2 to carbohydrate, and the cycle gives plants the metabolic flexibility to provide for this.

9.3 Patterns of electron and proton transport in photosynthetic bacteria

Successful techniques for elucidating pathways of electron and proton transport in bacterial photosynthesis have been the same as those used with green-plant tissues, as listed in the last section. In one sense the bacteria are intrinsically simpler, having only one kind of photosystem. This simplicity is offset in many types of photosynthetic bacteria by the complexity introduced by respiration. In contrast to plants, the sites and components of respiratory activity are not well separated from the photosynthetic pathways in bacteria, and the experimenter must learn how to discriminate between these activities and the compounds involved in them. A clear advantage in studying the bacteria is the possibility of isolating the reaction centers and observing their interactions with membrane components such as quinones and cytochromes. One must be careful, of course, in extrapolating these observations to the behavior of the reaction centers in situ.

The membranes of photosynthetic bacteria contain a variety of electron carriers, especially ubiquinone, cytochromes of the b and c types,[11] and Fe—S centers. Some of these carriers have been placed with confidence in a scheme of photosynthetic electron transport, and some have not. An outline of electron transport in purple photosynthetic bacteria is shown in Fig. 9.8, a counterpart of the "green plant" scheme of Fig. 9.5. The gaps in this outline signify areas of maximum ignorance. Until recently there was little hesitation in placing a cytochrome of the b type in a linear sequence $UQ \longrightarrow$ Cyt $b \longrightarrow$ Cyt c, just as we do not hesitate to place Cyt 563 between FD and PQ, and Cyt f between PQ and P700, in the scheme for green plants. We shall see, however, that there are obstacles to this simple assignment.

In plants the reduction of $NADP^+$ is effected directly from the strong

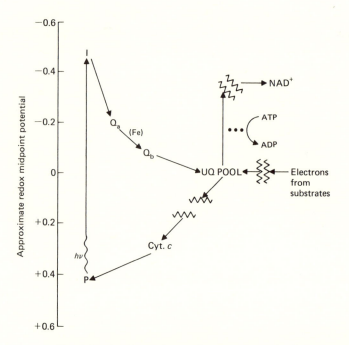

Fig. 9.8. A scheme for cyclic and noncyclic electron transport in purple photosynthetic bacteria, showing the approximate E_m values of the components. The gaps indicate areas of relative ignorance, where cytochromes of the b type and Fe—S centers are likely to participate.

reductants generated by Photosystem 1, and in the green sulfur bacteria a similar mechanism appears to bring about the reduction of NAD^+, the bacterial counterpart of $NADP^+$. In contrast, the reduction of NAD^+ in purple bacteria requires an input of energy in order to raise electrons from the redox level of the UQ pool to that of $NAD^+/NADH$ (E_m = -0.35 V). This energy can come from the hydrolysis of ATP, but it comes more directly from the "high energy state," denoted ⊖, that normally leads to ATP formation.[12] The antibiotic oligomycin prevents the utilization of ⊖ to make ATP, but it does not inhibit the light-induced reduction of NAD^+ in purple bacteria. This reduction *is* inhibited by uncouplers, agents that dissipate the energy of ⊖ before it can be used. The reduction of NAD^+ might involve one or more of the Fe—S centers, which have been detected by ESR but whose functions are uncertain.

When needed, electrons can be introduced from external substrates by way of specific substrate dehydrogenase enzymes. These electrons probably enter the UQ pool, again by largely undefined pathways.

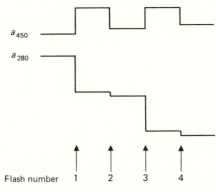

Fig. 9.9. Absorbance changes at 450 and 280 nm, showing the pattern of formation and disappearance of semiquinone and the accumulation of fully reduced quinone, in reaction centers of *Rp. sphaeroides* endowed with a pool of ubiquinone and secondary electron donor. (See discussion in the text.)

The flow of electrons from the reaction center into the UQ pool, and the attendant binding of H^+, has been studied with isolated reaction centers of *Rhodopseudomonas sphaeroides* as well as in chromatophores and cells of purple bacteria. By differential extraction and replacement of quinones, isolated reaction centers can be prepared with no quinone, with Q_a only, with Q_a and Q_b, or with the full complement of Q_a, Q_b, and a UQ pool of arbitrary size (in the living cell there are about 5 to 10 molecules of UQ per reaction center). In reaction centers endowed with a full complement of UQ, A. Vermeglio and C. A. Wraight[13] independently discovered a pattern similar to that found earlier in chloroplasts: the formation and disappearance of the semiquinone on alternate flashes of light. The conversion of UQ to $UQ^{-\cdot}$ was signaled by an increase of absorbance at 450 nm and a decrease at 280 nm; in the conversion of $UQ^{-\cdot}$ to UQ^{2-} or UQH_2 the absorbance at 280 nm did not change, and the absorbance at 450 nm returned to its original level (Fig. 9.9). In these experiments the reaction centers were provided with an excess of an electron donor in order to re-reduce P^+ rapidly after each flash. Later studies showed that at pH below 7 two H^+ are bound on even-numbered flashes and none on odd-numbered flashes.

These results are consistent with the mechanism described in the last section for PQ in chloroplasts, as shown by Reactions 9.6 but with UQ in place of PQ. The transfer of 2H from $Q_b H_2$ to the pool after an even-numbered flash might be achieved simply by $Q_b H_2$ "falling off," with UQ from the pool taking its place. This possibility has not been tested directly (e.g., by means of ^{14}C-labeled UQ), but M. Y. Okamura and G. Feher have observed that Q_b is extracted from reaction centers more readily in a reducing environment.

As expected, reaction centers with only Q_a show the simple pattern of $Q_a \longrightarrow Q_a^-$ after each flash, with the Q_a^- discharging its electron between flashes (presumably to an oxidized donor molecule in the medium) and reverting to Q_a:

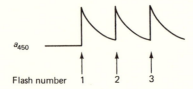

With Q_a and Q_b but no pool, Vermeglio observed the expected result:

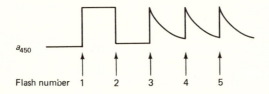

After the first two flashes, with Q_b fully reduced, the system behaves as if only Q_a is present.

Wraight showed that as the pH of the suspension of reaction centers is raised above 7, some H^+ is bound on the first flash, and at pH $\gtrsim 9$ the binding of H^+ is the same after every flash, even though the oscillating pattern of a_{450} persists. Wraight suggested that H^+ can be bound and released at a site near Q_a and Q_b, in coordination with the changing redox states of Q_a and Q_b. At low pH this site remains protonated and does not participate in H^+ binding; the mechanism of Reactions 9.6 is adequate. At high pH we might have

First flash:

$$\begin{array}{ccc} N^- & \xrightarrow{h\nu} & N^- & \xrightarrow{+H^+} & NH \\ Q_a Q_b & & Q_a^- Q_b & & Q_a Q_b^- \end{array}$$

Second flash:

$$\begin{array}{ccc} NH & \xrightarrow{h\nu} & NH & \xrightarrow{+H^+} & N^- \\ Q_a Q_b^- & & Q_a^- Q_b^- & & Q_a Q_b H_2 \end{array}$$

The plausibility of this model requires that the affinity of N^- for H^+ depend on the redox states of Q_a and Q_b.

The possibility that a site not involved directly in oxidation and reduction can bind and release protons in response to the states of nearby electron carriers had been invoked earlier by B. Chance, in speculating about H^+ binding in membranes. Chance referred to this as a "membrane Bohr effect."

The pattern of binary oscillations in the reduction of UQ and the binding of H^+ has been seen in chromatophores and intact cells as well as in isolated reaction centers, especially at higher ambient redox potential (greater than about $+0.2$ V). Some strains of *Rp. sphaeroides* show the oscillatory pattern more clearly than others. We must admit that the conditions that favor the observation of binary oscillations are not well defined, and the importance of the oscillatory mechanism in living cells remains a matter of conjecture. Most measurements with cells and chromatophores of purple bacteria have been made under conditions in which oscillations were not observed, a series of flashes giving the same result (binding of H^+ and the transient appearance of UQ^-) after each flash. We shall speculate about possible reasons for this later in this section.

In reaction centers Q_b is distinct from other UQ molecules in the pool. The most concrete evidence for this is that exposure to a high concentration of a detergent such as LDAO removes the pool and yields reaction centers that retain just Q_a and Q_b. The ESR signals of both Q_a^- and Q_b^- show the magnetic influence of the iron atom that is bound to the reaction center, suggesting that these quinones are close to each other and to the Fe atom. One would not expect $UQ^-\cdot$ in the pool to show a strong magnetic interaction with this specific Fe atom. Furthermore the formation of either Q_a^- or Q_b^- is attended by shifts of the absorption bands of Bpheo and Bchl in the reaction center, and the patterns of these shifts are distinctive and different for Q_a^- and Q_b^-. The band shifts that attend the reduction of Q_b to Q_b^- are the same whether or not a large pool of UQ is present, indicating again that Q_b^- has a specific location with respect to the reaction center. Exactly the same band shifts related to Q_a^- and to Q_b^- are seen in chromatophores; this is good evidence that both Q_a and Q_b exist as specific entities in the intact photosynthetic membrane, with properties distinct from those of other UQ molecules comprising a larger pool.

There are several possible reasons that cells and chromatophores exposed to consecutive flashes show oscillations of quinone reduction and H^+ uptake under some conditions and not under others. One possibility is that molecules of semiquinone, produced by many reaction centers in a single flash, can interact through a communal pool of UQ:

$$2UQ^-\cdot + 2H^+ \longrightarrow UQ + UQH_2$$

This would require that the electrons delivered from the reaction centers after a flash be confined to Q_b^- only under the conditions in which oscillations are observed. Under conditions that do not show oscillations, Q_b^- can release its single electron to the pool. No one has advanced a rational basis for such a change of electron mobility in response to slight changes in the chemical environment (redox potential) of the photosynthetic membrane.

A second possibility is that oscillations become unobservable because a mixed population of photosynthetic units has developed, with some in the state $Q_aQ_b^-$ and others in the state Q_aQ_b. With equal numbers in each state, the flash-induced reactions $Q_aQ_b \longrightarrow Q_aQ_b^-$ and $Q_aQ_b^- \longrightarrow Q_aQ_b^{2-} \longrightarrow Q_aQ_b$ nullify each other (this accounts for the relatively low amplitude of "plastoquinone oscillations" seen in green-plant tissues). A mixed population will arise during illumination because of the random way in which reaction centers receive and respond to light quanta; this accounts for the damping of oscillations in a long series of flashes. Several minutes of dark-adaptation are then needed, to allow $Q_aQ_b^-$ to relax to the state Q_aQ_b, before a new sequence of flashes will show maximal oscillations. This relaxation requires the loss of an electron from Q_b^- to its surroundings, and such a process should be slower at lower ambient redox potential. The reported absence of oscillations at low redox potential could be due simply to insufficient dark-adaptation prior to each set of flashes.

A third possibility is that Q_b^-, once formed, can receive a second electron rapidly from another molecule such as reduced cytochrome, and this represents the "normal" operation of the electron transport cycle. Oscillations can only be observed if Q_b^- formed in one flash is stable until the next flash is applied; this could come about if the cytochrome (or other electron donor) is in its oxidized state, and hence unable to give an electron to Q_b^-, when the flashes are presented. This explanation will be discussed in more detail later, in connection with Figs. 9.10 and 9.11.

Turning to the oxidizing side of the reaction center, the most immediate electron donor to P^+ in the living tissue is a cytochrome of the c type. In $Rp.$ *sphaeroides* there are two apparently identical molecules of "Cyt c_2" (E_m = +0.3 V; α band maximum at 551 nm in the reduced form) associated with each reaction center. A flash that generates P^+ will cause the oxidation of either of these two Cyt c_2 molecules with equal facility, and there has been no evidence to distinguish them in their reactions with other electron carriers in the membrane. We do not know why there are two for each reaction center; it could have to do with an interface between Cyt c_2 and UQ. $Rp.$ *sphaeroides* also contains three distinct cytochromes of the b type in roughly 1:3:2 ratio: Cyt b_{-90} (E_m -0.09 V; λ_{max} 564 nm), Cyt b_{50} (E_m +0.05 V; λ_{max} 560 nm), and Cyt b_{155} (E_m +0.155 V; λ_{max} 559 nm). The light-induced reactions of these cytochromes will be described later; only Cyt b_{50} has been shown convincingly to be in the mainstream of cyclic electron flow. The types and properties of light-reacting cytochromes in $Rp.$ *capsulata* and $Rs.$ *rubrum* are closely similar to those in $Rp.$ *sphaeroides.*

A second major category of purple bacteria in terms of cytochrome content and function includes $Rp.$ *viridis*, $Rp.$ *gelatinosa*, and *Chr.* *vinosum.* In these organisms little or no Cyt b has been associated convincingly with the light reactions. Instead, but not necessarily fulfilling comparable functions, there are low-potential cytochromes of the c type (E_m in the range 0 to +0.1 V) as well as c-type cytochromes with E_m near +0.3 V. Properties of these cytochromes are presented in Table 9.3. Reaction centers isolated from

Table 9.3. *Properties of high- and low-potential cytochromes of the c type in three species of purple bacteria*

Organism	Cyt c, designated by λ_{max} of α band	Approximate E_m (V)
Rhodopseudomonas viridis	Cyt 558	+0.33
	Cyt 552	0.0
Rhodopseudomonas gelatinosa	Cyt 553	+0.28
	Cyt 548	+0.13
Chromatium vinosum	Cyt 555	+0.34
	Cyt 553	0.0

Rp. viridis and *Chr. vinosum* retain firmly bound cytochromes of both the high- and low-potential c types; for *Rp. viridis* the ratio is two Cyt 558 and three Cyt 552 per reaction center. In *Chr. vinosum* there are two Cyt 555 and two Cyt 553 for each reaction center.

The two categories of purple bacteria, exemplified by *Rp. sphaeroides*, *Rp. capsulata*, and *Rs. rubrum* (first category) and by *Rp. viridis, Rp. gelatinosa*, and *Chr. vinosum* (second category), differ strikingly in the ways that their c-type cytochromes are associated with the membrane and react with P^+. In those of the first category the Cyt c_2 is associated loosely with the membrane on the periplasmic side. Much of it is washed away when the cells are broken, but some is trapped inside the chromatophores:

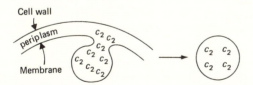

In contrast, the cytochromes c of the second category are integral components of the membrane, and in some cases remain firmly bound to the isolated reaction centers.

Cytochromes c_2 in organisms of the first category donate electrons to P^+ in times of the order of tens of microseconds at room temperature, and not at all at liquid-nitrogen temperature (77 K). The cytochromes c in the second category donate to P^+ in about 2 μsec or less at room temperature. The low-potential ones, and some of the high-potential ones, can donate at liquid-nitrogen temperature. The reaction times for electron donation from cytochromes to P^+, as functions of temperature, have been analyzed in terms of vibronically assisted quantum-mechanical electron tunneling (see J. Jortner, suggested reading, Chapter 8).

The low-potential cytochromes c in the second category, and Cyt b_{155} in the first, have been implicated in the noncyclic introduction of electrons from external substrates. The low-potential c types are oxidized rapidly (in microseconds) by P^+, but their re-reduction requires many seconds, making them unlikely candidates for carriers in a rapid cycle. Cyt b_{155} is oxidized, giving electrons to P^+ by way of Cyt c_2, within a few milliseconds after a flash. Re-reduction of Cyt b_{155} requires about a second, again too slow for a rapid cycle but adequate for the less frequent introduction of electrons from external substrates. In contrast to the slow re-reduction of these cytochromes, the high-potential c types, including Cyt c_2, are re-reduced in milliseconds following their oxidation. Thus the rate of electron flow through the high-potential cytochromes (cyclic path) can be about 1000 times the rate of flow through the low-potential cytochromes (hypothetical noncyclic path) in continuous light. This is consistent with the general observation that cyclic electron flow, coupled to ATP formation, is the principal light-driven activity in purple bacteria. Electrons from external substrates need to be brought in only occasionally, to preserve an overall metabolic redox balance while carbon compounds are being assimilated. If the low-potential cytochromes bring in electrons from external substrates, these electrons are probably deposited in the UQ pool, following the path

$$\text{"substrate} \longrightarrow \cdots \longrightarrow \text{Cyt } c \text{ (low-pot.)} \longrightarrow \text{P} \xrightarrow{h\nu} \text{UQ"}$$

or

$$\text{"substrate} \longrightarrow \cdots \longrightarrow \text{Cyt } b_{155} \longrightarrow \text{Cyt } c_2 \longrightarrow \text{P} \xrightarrow{h\nu} \text{UQ."}$$

The various Fe$-$S centers that have been detected in purple bacteria, with E_m values ranging from about $+0.3$ to -0.4 V, could have a variety of roles, none of them documented: in the cyclic and noncyclic light-driven pathways of electron transfer, in the energy-linked reduction of NAD^+, and in respiration.

The gap in Fig. 9.8 between UQ and Cyt c, and the related question of the reactions of Cyt b, remain to be considered. The relationships between b cytochromes and other electron and proton carriers in purple bacteria have been studied most extensively with chromatophores of *Rp. sphaeroides*, especially by P. L. Dutton and collaborators (see the suggested reading). The study of light-induced reactions of these cytochromes has been difficult and confusing because of their multiplicity and their relative remoteness from the primary photochemical reaction. The main approach has been to measure flash-induced absorbance changes as functions of the ambient redox potential, pH, and presence or absence of various inhibitors, and to try to dissect these observations to reveal the reactions of specific cytochromes.

Within the first 200 μsec after a flash, chromatophores of *Rp. sphaeroides* under "physiological" conditions show the oxidation of Cyt c_2 (by P^+) and the uptake of H^+, presumably by reduced UQ. During the next few milliseconds one sees the re-reduction of Cyt c_2, the oxidation of Cyt b_{155}, and a wave of reduction followed by reoxidation of Cyt b_{50}. Cyt b_{-90} also engages in light reactions on this time scale, but the pattern of its reactions is obscure. The Cyt b_{155} is re-reduced slowly, in about a second.

The reoxidation of Cyt b_{50}, after its flash-induced reduction, has kinetics similar to those of the re-reduction of Cyt c_2, but the component that re-reduces Cyt c_2 is apparently not Cyt b_{50}. A redox titration of this hypothetical component, measuring the kinetics of Cyt c_2 re-reduction as a function of ambient redox potential, gives $E_m = +0.15$ V and $n = 2$, possibly reflecting a "special" UQ (the main UQ pool shows $E_m = +0.05$ V). This hypothetical donor of electrons to Cyt c_2 has been called Z.

Antimycin a is an inhibitor that blocks both the reoxidation of Cyt b_{50} and the re-reduction of Cyt c_2; perhaps it acts on Z. In its presence, a sequence of flashes causes the irreversible reduction of Cyt b_{50} (essentially complete after the first flash) and the irreversible oxidation of Cyt c_2 (nearly complete after two flashes; there are two Cyt c_2 for each reaction center). Each flash leads to the binding of one H^+ per reaction center with half-time about 150 μsec, as long as electrons are available to re-reduce P^+ and prepare it for the next photoact. This H^+ binding has been attributed to the protonation of $UQ^-\cdot$, but it could be due to a "membrane Bohr effect" (see the earlier discussion of this effect on page 216).

In the absence of antimycin a, chromatophores of *Rp. sphaeroides* bind two H^+ for every electron driven photochemically through the reaction center and around the cycle. This remarkable observation has been a major factor in stimulating thought about cycles more complicated than the simple UQ $\longrightarrow$ Cyt $b \longrightarrow$ Cyt $c \longrightarrow$ P870 sequence that had been visualized earlier. UQ is not the only potential proton carrier; Cyt b_{50} also has this capability. Below pH 7.4, the E_m of Cyt b_{50} varies with pH according to $\Delta E_m/\Delta pH = -0.06$ V. This implies that the oxidation and reduction of Cyt b_{50} below pH 7.4 should be described as

$$\text{Cyt } b_{50}(\text{ox}) + e^- + H^+ \longrightarrow \text{Cyt } b_{50}(\text{red})\cdot H^+$$

and permits us to draw schemes in which Cyt b_{50} helps to carry H^+ across the membrane.

In all of these observations on cytochrome reactions and H^+ binding in chromatophores of *Rp. sphaeroides*, the conditions were such that binary oscillations of semiquinone formation and disappearance on consecutive flashes were not reported. Under conditions that reveal these oscillations, describable vaguely as "high redox potential," H^+ is bound only on alternate flashes. The stoichiometry has not been determined accurately, but the most likely value is two H^+ per reaction center on each even-numbered flash and none on odd-numbered flashes.

Attempts have been made, especially by P. L. Dutton and by A. R. Crofts, to draw schemes for electron and proton transport that can account for most

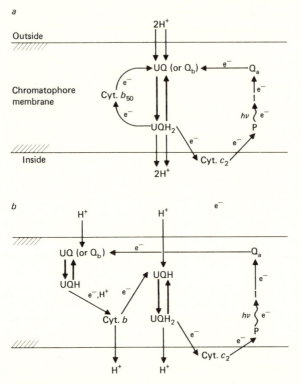

Fig. 9.10. Two hypothetical outlines of cyclic electron and H^+ transport in *Rp. sphaeroides;* condensed or slightly modified versions of schemes proposed by P. L. Dutton (*a*) and A. R. Crofts (*b*) (see Dutton and Prince, 1978a, and Crofts and Wood, suggested reading for this chapter). Components such as Cyt b_{-90} and Z, the quinone-like hypothetical electron donor to Cyt c_2, have been omitted.

of the observations made with *Rp. sphaeroides*, many details of which have been omitted from this account. Two contrasting schemes, one favored by Dutton and the other by Crofts, are shown in skeletal form (with several components omitted for clarity) in Fig. 9.10. Both versions provide for two H^+ translocated for every electron driven through the reaction center. The binary quinone oscillations might occur only when certain components are inoperative. For example, in the upper scheme (*a*), if the Cyt b_{50} is kept permanently oxidized by external conditions, so that it is unable to deliver electrons to UQ, a truncated cycle might operate as shown in Fig. 9.11. In the truncated version two consecutive turnovers are required in order to render UQ fully reduced and able to bind $2H^+$.

The schemes of Figs. 9.10 and 9.11 are presented merely as examples of working hypotheses, subject to frequent revision or outright rejection. They

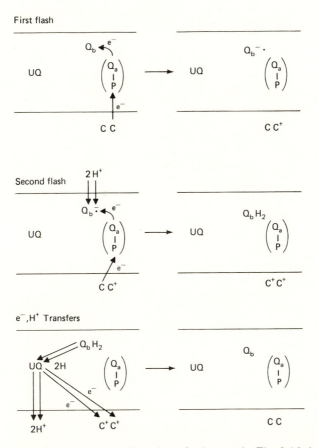

Fig. 9.11. A truncated version of scheme *a* in Fig. 9.10, in which the presence of semiquinone and the translocation of H^+ show binary oscillations on consecutive flashes. C represents cytochrome c_2.

illustrate an interesting possibility for physiological flexibility: Under anaerobic conditions the bacteria can translocate two H^+ per quantum and make a commensurate amount of ATP. When the bacteria find themselves in an aerobic environment that oxidizes some of the electron carriers and disrupts the cyclic pattern, they can still maintain themselves with a truncated cycle, translocating H^+ and making ATP at half the former efficiency. Alternatively, the binary quinone oscillations might be intriguing laboratory artifacts that seldom occur in nature.

Many of the foregoing observations made with *Rp. sphaeroides* have been confirmed, or made independently, with chromatophores from *Rp. capsulata* and *Rs. rubrum*. As little as we understand the details of electron and proton transport in these creatures, we know even less about the category that includes *Rp. viridis*, *Rp. gelatinosa*, and *Chr. vinosum*. Perhaps in these organisms, a simple cycle of the form

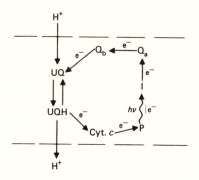

or perhaps a double shuttle such as

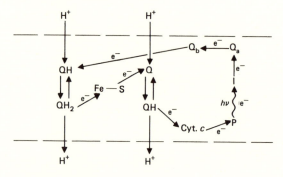

is close to reality.

The green sulfur bacteria have been studied far less than the purple bacteria, and our ignorance of photosynthetic electron and proton transport in the former is correspondingly more profound. A tentative outline based on fragments of evidence is shown in Fig. 9.12. The primary photochemical donor P840 (E_m = +0.3 V) delivers electrons to an acceptor, probably an Fe—S center of E_m = -0.55 V (see the last part of Section 8.1). Other Fe—S centers with E_m values of +0.16, -0.02, and -0.17 V have been detected by ESR, but their functions are unknown.

Electrons driven through the reaction center can reduce NAD^+ by way of a soluble ferredoxin and FD—NAD reductase. In this reaction the native FD can be replaced by FD from *Chr. vinosum* (E_m = -0.49 V) and the reductase can be replaced by FD—NADP reductase from spinach. This noncyclic path is fed by electrons from substrates, notably sulfide (S^{2-}) and thiosulfate ($S_2O_3^{2-}$). These electrons enter the pathway to the reaction center, probably by way of Cyt b ⟶ Cyt 555 ⟶ P840. The path from substrate to Cyt b includes a specific c-type cytochrome, Cyt 551 (E_m = +0.13 V) for $S_2O_3^{2-}$

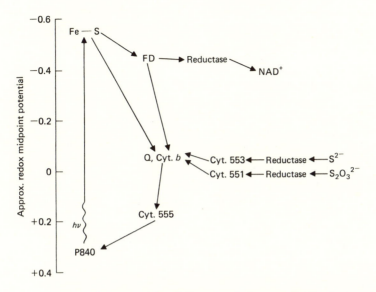

Fig. 9.12. A provisional diagram of photosynthetic electron transport in green sulfur bacteria. Q represents a quinone pool of uncertain significance in photosynthesis.

and Cyt 553 (E_m = +0.10) for S^{2-}. Substrates are linked to these c cytochromes by specific reductases.

In nature these bacteria typically reduce CO_2 to cell materials, using electrons from inorganic substrates such as H_2S. This implies that the noncyclic pathway to NAD^+ is a major one, but a cyclic pathway may also be needed in order to generate ATP. The cyclic path shown in Fig. 9.12, through FD, Cyt b and/or quinone, and Cyt 555 is at least consistent with available evidence. The quinone of green sulfur bacteria is about half menaquinone and half a variant called *Chlorobium* quinone; we know almost nothing about its possible light-induced oxidation and reduction.

The Cyt b (E_m = -0.1 V and λ_{max} = 564 nm) can be either reduced or oxidized by light in subcellular preparations, depending on the ambient redox potential. Its reduction is stimulated by antimycin a, suggesting (in analogy with purple bacteria) that it normally delivers electrons to the c-type Cyt 555. There are two Cyt 555 per reaction center; its E_m is +0.2 V and it is oxidized rapidly by light, probably interacting directly with P^+.

Detailed pathways and mechanisms of proton translocation in the green sulfur bacteria remain to be elucidated. The schemes of Figs. 9.5, 9.6, 9.8, and 9.12 appear simple when compared with Fig. 9.10, but this does not necessarily mean that the purple bacteria use more intricate pathways of H^+ transport. With further study the seemingly simpler networks of electron flow shown for green plants and green sulfur bacteria may have to be made more complicated in order to account for the details of proton translocation.

SUGGESTED READINGS

Joliot, P., Barbieri, G., and Chabaud, R. (1969). Un nouveau modèle des centres photochimiques du Système II. *Photochem. Photobiol. 10*, 309–29.

Kok, B., Forbush, B., and McGloin, M. (1970). Cooperation of charges in photosynthetic oxygen evolution–I. A linear four step mechanism. *Photochem. Photobiol. 11*, 457–75.

Joliot, P., and Kok, B. (1975). Oxygen evolution in photosynthesis. In *Bioenergetics of Photosynthesis*, Govindjee, ed., Chapter 8, pp. 387–412. Academic Press, New York.

van Best, J. A., and Mathis, P. (1978). Kinetics of the reduction of the primary electron donor of Photosystem II in spinach chloroplasts and in *Chlorella* cells in the microsecond and nanosecond time range following flash excitation. *Biochim. Biophys. Acta 503*, 178–88.

Blankenship, R. E., McGuire, A., and Sauer, K. (1977). Rise time of EPR Signal II_{vf} in chloroplast Photosystem 2. *Biochim. Biophys. Acta 459*, 617–19.

Junge, W., Aüslander, W., McGeer, A. J., and Runge, T. (1979). The buffering capacity of the internal phase of thylakoids and the magnitude of the pH changes inside under flashing light. *Biochim. Biophys. Acta 546*, 121–41.

Saphon, S., and Crofts, A. R. (1977). Protolytic reactions in Photosystem II: A new model for the release of protons accompanying the photooxidation of water. *Z. Naturforsch. 32c*, 617–26.

Amesz, J., and Duysens, L. N. M. (1977). Primary and associated reactions of System II. In *Primary Processes of Photosynthesis*, J. Barber, ed., Chapter 4 (pp. 149–85). Elsevier, Amsterdam.

Trebst, A. (1974). Energy conservation in photosynthetic electron transport of chloroplasts. *Annu. Rev. Plant Physiol. 25*, 423–58.

Bouges-Bocquet, B. (1973). Electron transfer between the two photosystems in spinach chloroplasts. *Biochim. Biophys. Acta 314*, 250–6.

Pulles, M. P. J., Van Gorkom, H. J., and Willemsen, J. G. (1976). Absorbance changes due to the charge-accumulating species in System 2 of photosynthesis. *Biochim. Biophys. Acta 449*, 536–40.

Witt, H. T. (1975). Primary acts of energy conservation in the functional membrane of photosynthesis. In *Bioenergetics of Photosynthesis*, Govindjee, ed., Chapter 10, pp. 493–554. Academic Press, New York.

Bouges-Bocquet, B. (1977). Cytochrome *f* and plastocyanin kinetics in *Chlorella pyrenoidosa*. I. Oxidation kinetics after a flash. *Biochim. Biophys. Acta 462*, 362–70.

Haehnel, W. (1977). Electron transport between plastoquinone and chlorophyll A_1 in chloroplasts. II. Reaction kinetics and the functioning of plastocyanin in situ. *Biochim. Biophys. Acta 459*, 418–41.

Katoh, S. (1971). Plastocyanin. *Methods Enzymol. 23A*, 408–13.

Malkin, R., Knaff, D. B., and Bearden, A. J. (1973). The oxidation–reduction potential of membrane-bound chloroplast plastocyanin and cytochrome *f*. *Biochim. Biophys. Acta 305*, 675–8.

Böger, P. (1978). Some properties of plastocyanin and its function in algal chloroplasts. In *Proc. 4th Internatl. Cong. Photosynthesis*, D. O. Hall, J. Coombs, and T. W. Goodwin, eds., pp. 755-64. The Biochemical Society, London.

Malkin, R., and Bearden, A. J. (1978). Electron paramagnetic resonance studies of plastocyanin in the photosynthetic electron transport chain. In *Proc. 4th Intl. Cong. Photosynthesis*, D. O. Hall, J. Coombs, and T. W. Goodwin, eds., pp. 787-91. The Biochemical Society, London.

Telfer, A., and Evans, M. C. W. (1972). Evidence for chemiosmotic coupling of electron transport to ATP synthesis in spinach chloroplasts. *Biochim. Biophys. Acta 256*, 625-37.

Junge, W. (1977). Membrane potentials in photosynthesis. *Ann. Rev. Plant Physiol. 28*, 503-36.

Knaff, D. B. (1978). Reducing potentials and the pathway of NAD^+ reduction. In *The Photosynthetic Bacteria*, R. K. Clayton and W. R. Sistrom, eds., Chapter 32, pp. 629-40. Plenum, New York.

de Grooth, B. G., van Grondelle, R., Romijn, J. C., and Pulles, M. P. J. (1978). The mechanism of reduction of the ubiquinone pool in photosynthetic bacteria at different redox potentials. *Biochim. Biophys. Acta 503*, 480-90.

Wraight, C. A. (1979). Electron acceptors of bacterial photosynthetic reaction centers. II. H^+ binding coupled to secondary electron transfer in the quinone acceptor complex. *Biochim. Biophys. Acta 548*, 309-27.

Dutton, P. L., and Prince, R. C. (1978a). Reaction center driven cytochrome interactions in electron and proton translocation and energy coupling. In *The Photosynthetic Bacteria*, R. K. Clayton and W. R. Sistrom, eds., Chapter 28, pp. 525-70. Plenum, New York.

Dutton, P. L., and Prince, R. C. (1978b). Equilibrium and disequilibrium in the ubiquinone-cytochrome b-c_2 oxidoreductase of *Rhodopseudomonas sphaeroides*. *FEBS Lett. 91*, 15-20.

Knaff, D. B., and Buchanan, B. B. (1975). Cytochrome b and photosynthetic sulfur bacteria. *Biochim. Biophys. Acta 376*, 549-60.

Crofts, A. R., and Wood, P. M. (1978). Photosynthetic electron-transport chains of plants and bacteria and their roles as proton pumps. *Curr. Top. Bioenerg. 7*, 175-244.

Part IV

THE FORMATION OF ATP AND THE ASSIMILATION OF CARBON DIOXIDE

We come now to the more peripheral topics of ATP formation and CO_2 assimilation. The central problem in ATP formation is "How is the redox energy of electron transport transmuted into the anhydride bond energy of ATP?" We shall consider three distinct hypotheses and focus our attention on one, the chemiosmotic hypothesis that was the basis of Peter Mitchell's Nobel Prize in chemistry in 1978. The chemiosmotic hypothesis provides the best model available for understanding the multifarious phenomena of light-driven ion (primarily H^+) transport in photosynthetic membranes, and the interrelated formation of ATP. We shall examine the methods used to probe electrochemical gradients in or across the membranes, and consider the patterns of electron and ion movements that generate these gradients. The kind of electrochemical gradient that is correlated with ATP formation is the gradient in proton potential. This includes two terms: the gradient in H^+ concentration (ΔpH) across the membrane, and any gradient of electric potential across the membrane ($\Delta\psi$) that would tend to drive H^+ down its concentration gradient. Figure 10.1 shows in which direction the gradient is formed by light. We shall go on to examine the ATP-forming enzyme and its

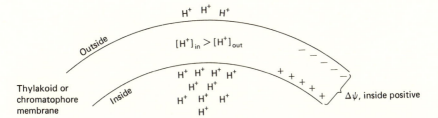

Fig. 10.1. Two components of the light-induced gradient in proton potential across the thylakoid membrane. (See text for discussion.)

responses to the protonmotive gradient, and finally consider the overall energy efficiencies of plant and bacterial photosyntheses.

The treatment of carbon assimilation will be superficial; this topic is far from the central considerations of this book. We shall examine two pathways in plants: the reductive pentose cycle and the phosphoenolpyruvate carboxylation path.

10 Electrochemical gradients and the formation of ATP

10.1 Hypotheses for the coupling of electron transport to ATP formation

The chemical structure of ATP is shown in Fig. 1.2. The structure of ADP can be abbreviated as

$$
\text{adenine-ribose} - O - \underset{\underset{OH}{|}}{\overset{\overset{O}{\|}}{P}} - O - \underset{\underset{OH}{|}}{\overset{\overset{O}{\|}}{P}} - OH
$$

and if all but the right-hand —OH group is represented by R, the conversion of ADP to ATP can be written

$$
\underset{\text{(ADP)}}{R-OH} + \underset{\underset{OH}{|}}{\overset{\overset{O}{\|}}{\underset{\text{(o-phosphoric acid)}}{HO-P-OH}}} \longrightarrow \underset{\underset{OH}{|}}{\overset{\overset{O}{\|}}{\underset{\text{(ATP)}}{R-O-P-OH}}} + H_2O \qquad (10.1)
$$

At physiological pH some of the —OH groups are ionized and the reaction should be written

$$
R-O^- + \underset{\underset{O^-}{|}}{\overset{\overset{O}{\|}}{HO-P-O^-}} + H^+ \longrightarrow R-O-\underset{\underset{O^-}{|}}{\overset{\overset{O}{\|}}{P}}-O^- + H_2O \qquad (10.2)
$$

The process is one of dehydration, the reverse of hydrolysis. In an aqueous environment the newly formed anhydride bond is a high energy bond.

Redox energy is basically the energy of separated electric charges (Section 4.3). In photosynthesis the photoact generates this redox energy, and it becomes available in the subsequent "downhill" flow of electrons through carriers of ever more positive redox potential. The question of how this energy is converted to anhydride bond energy in ATP, in respiration as well as in photosynthesis, has been the preoccupation of many scientists for decades.

231

A basic aspect of this conversion of energy is the phenomenon of coupling: The rate of energy-yielding electron transport is constrained by the rate at which energy can be used in forming ATP. In healthy living cells, if the conversion of ADP to ATP is slow because (for example) too little ADP is available, the rate of electron transport is slowed correspondingly. The constraint is lifted by agents called uncouplers, which divert the flow of energy and allow electron transport to run freely but unproductively. All coupling hypotheses invoke a high energy intermediate(s), symbolized $\ominus$, that connects electron transport and ATP formation. The accumulation of $\ominus$ limits the rate of electron transport. Uncouplers act as if they dissipate $\ominus$:

$$\text{Electron transport} \longrightarrow \ominus \longrightarrow \text{ATP formation}$$
$$\downarrow \text{(uncoupler)}$$
$$\text{heat}$$

The experimenter can thus manipulate the process at three levels: (1) Inhibitors of the ATP-forming enzyme, including the lack of substrates, attenuate both ATP formation and electron transport while allowing $\ominus$ to accumulate. (2) Uncouplers dissipate $\ominus$ and stop ATP formation while allowing electron transport to proceed at maximal rates. (3) Direct inhibition of electron transport stops the entire process.

The distinction between different coupling hypotheses lies in the presumed nature of $\ominus$.

The conformational coupling hypothesis holds that the transport of electrons induces stresses that change the conformations of macromolecules; in consequence of this the equilibrium between hydrolysis and dehydration of one or more systems (including ADP $\longleftrightarrow$ ATP) is shifted to favor ATP formation. This hypothesis is so lacking in specific definition that it is hard to adduce evidence for or against it, as either an alternative or an embellishment to other hypotheses.

In the chemical coupling hypothesis, $\ominus$ is identified as a molecule or molecular complex that combines the chemical properties of oxidation-reduction and hydrolysis-dehydration. This complex has the property that the affinity for H_2O is changed by a change of the redox state of the electron-carrying part, and the E_m of the latter is dependent on the state of hydration. The chemical coupling hypothesis was successful in accounting for the formation of ATP linked to the fermentation of sugars and other organic compounds through soluble components. In these processes the events that correspond to oxidation-reduction and hydrolysis-dehydration are intertwined and not always easy to discriminate, as when the oxidation of a substrate involves incorporation of an oxygen atom originating in phosphate or water. An example is the oxidation of glyceraldehyde-3-phosphate, linked to the reduction of NAD^+ and the conversion of ADP to ATP:

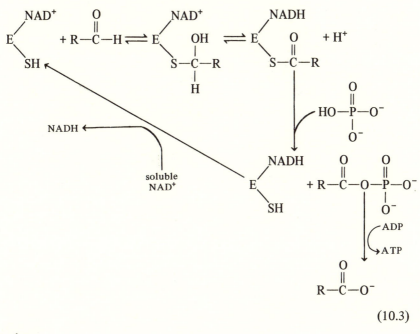

(10.3)

where

E = enzyme with cysteine —SH group

$$R-\overset{\overset{\displaystyle O}{\|}}{C}-H = \text{glyceraldehyde-3-phosphate,}$$

$$^{-}O-\overset{\overset{\displaystyle O}{\|}}{\underset{\underset{\displaystyle O^{-}}{|}}{P}}-O-CH_2-CHOH-\overset{\overset{\displaystyle O}{\|}}{C}-H$$

$$R-\overset{\overset{\displaystyle O}{\|}}{C}-O^{-} = \text{3-phosphoglyceric acid}$$

The steps in Reaction 10.3 are obligately connected but reversible; with abundant ATP the reverse process can become thermodynamically favored. The energy of ATP can then drive the flow of electrons "uphill," against the thermodynamic gradient. This is a model for energy (ATP)-linked reductions. The action of uncouplers is to short-circuit one or more steps in the sequence.

By about 1960 the chemical coupling hypothesis was beset by severe difficulties when applied to oxidative (respiration-linked) phosphorylation in mitochondria and to photosynthetic phosphorylation. First, there was the failure of all efforts to find and identify the hypothetical coupling intermediates. Second, there was the general observation that both oxidative and photosynthetic phosphorylation require the intactness of vesicular membranous structures that separate interior and exterior aqueous regions. This does not weigh specifically against the chemical coupling hypothesis; it simply seems irrelevant and gratuitous, and calls for some kind of explanation. Third, the agents that act as uncouplers have a degree of chemical diversity that is hard to reconcile with a mechanism of intervention with specific chemical coupling events. Uncouplers include detergents, amines including ammonia, and diverse organic compounds.

Recognizing these problems, and drawing on earlier speculations about such physiological processes as the secretion of HCl by the gastric mucosa and the transport of ions by roots and yeasts, P. Mitchell developed a chemiosmotic hypothesis for the coupling of electron transport to ATP formation. The very problems that weakened the chemical coupling hypothesis gave strength to the chemiosmotic hypothesis.

Mitchell proposed that the coupling intermediate $\ominus$ is a thermodynamic state equivalent to a proton concentration cell. Imagine a vessel with two compartments containing solutions of different concentrations of HCl, the higher concentration on the left:

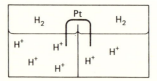

Let us add to this system by placing both compartments under a common atmosphere of H_2, allowing the system to come to equilibrium, and then connecting the two compartments with a platinum wire. Leaving the Cl^- out of the drawing,

In each compartment the H^+, the H_2, and electrons in the wire will engage in the reaction $2H^+ + 2e^- \longleftrightarrow H_2$. With a higher concentration of H^+ in the left compartment, the conversion of $2H^+ + 2e^-$ to H_2 will be more rapid than in the right compartment. The net result will be a flow of electrons through the wire from right to left and the transfer of H^+ (by way of H_2) from the left compartment to the right one. If neither anion in the solution, Cl^- or OH^-, contributes to this electrochemistry by donating electrons, the flow of current will soon cease because the solutions will have lost their electrical neutrality. The compartment on the left, having lost H^+, acquires an excess of negative charge (excess Cl^-) that opposes the influx of electrons from the wire, and vice versa on the right. Now if the partition were made selectively permeable to Cl^- (but not to H^+), electrical neutrality could be preserved by the passive diffusion of Cl^- from left to right, and the flow of current would continue until the concentration of HCl was equal on each side. The voltage developed by this proton battery is given by a Nernst equation, E (volts) = 0.06 log ($[H^+]_L / [H^+]_R$) = 0.06 (ΔpH), where ΔpH is the difference in pH on the two sides. The battery can deliver energy, as electrical work, equal to the integrated product of voltage and current until it has run down.

In the living analog of this proton battery, as conceived in the chemiosmotic hypothesis, the two compartments are the inside and the outside of a vesicle, and the partition is a membrane. The combination of H_2 gas and Pt wire is replaced by an ATP-forming enzyme (ATPase) situated in the membrane. The flow of current is a flow of H^+ through the enzyme, and the work performed (by mechanisms still poorly understood, to be discussed in Section 10.3) is the conversion of ADP to ATP. In the thylakoids of chloroplasts the battery is charged by the light-driven translocation of H^+ from outside to inside (see Fig. 9.6), and electrical neutrality is preserved by a compensating passive influx of Cl^- and efflux of Mg^{2+}. These passive ion fluxes are reversed in the discharge of the battery, while protons move out through the ATPase and ATP is formed. As in the inorganic analog, the proton potential equals 0.06 V for each unit of pH difference between inside and outside.

An outline of this conception for thylakoids and a contrasting one for the chromatophores of purple bacteria are shown in Fig. 10.2. In chromatophores the passive diffusion of ions is relatively slow. As a result the light-induced inward pumping of H^+ becomes limited by the excess of positive charge inside. There is excess H^+ inside the membrane, and excess OH^- on the outside. The resulting electric potential across the membrane, $\Delta\psi$, resists further inward pumping of H^+, but it also tends to drive H^+ out through the ATPase. To the extent that ions do move passively so as to neutralize some of the imbalance of charge, a certain amount of membrane potential is converted to an equivalent amount of pH differential.[1] Comparing the two modes of energy storage, one unit of pH difference across the membrane is equivalent to 0.06 V of membrane potential. Either currency is accepted by the ATPase. These two components of proton potential make up what Mitchell called the protonmotive force, pmf:

$$\text{pmf (volts)} = 0.06 \, (\Delta\text{pH}) + \Delta\psi \tag{10.4}$$

The action of uncouplers is readily understood in terms of the chemiosmotic hypothesis; they are agents that permit the passive movements of H^+

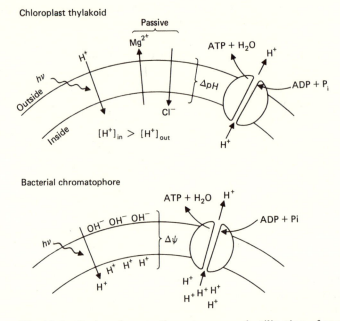

Fig. 10.2. Cycles of generation, storage, and utilization of energy in the form of proton potential, in chloroplast thylakoids and in chromatophores of purple bacteria.

and other ions so as to wash out the pH differential and/or the membrane potential. The actions of specific uncouplers will be described in Section 10.2.

The reverse process in which the hydrolysis of ATP drives electrons from more positive to more negative redox potential can be understood by a detailed scrutiny of the linked reactions, invoking the mass-action principle. Hydrolysis of ATP drives protons to the inside of the thylakoid or chromatophore, and the higher H^+ concentration shifts the equilibria of redox reactions that involve protons: A reaction of the type $A + e^- + H^+ \longleftrightarrow AH$ is driven toward the right.

The direction of H^+ movement is the same in chloroplast thylakoids and in bacterial chromatophores; for intact cells of purple bacteria the reverse holds. Light drives H^+ outward through the cytoplasmic membrane and its invaginations, from the cytoplasm to the periplasm. Similarly, the direction of H^+ movement driven by respiration in mitochondria is from inside to outside the mitochondrial membrane, and, as with the purple bacteria, the disruption of this membrane produces fragments that form into "inside-out" vesicles, analogous to chromatophores. These topological similarities are consistent with other similarities between phosphorylation in mitochondria and purple bacteria, such as the nature of the ATPase and its responses to specific inhibitors, and with evolutionary speculations about the bacterial origin of mitochondria.

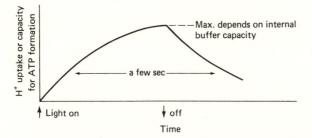

Fig. 10.3. Kinetics of proton uptake *or* capacity for ATP formation of chloroplasts during and after illumination, showing the identical relationship.

The chemiosmotic hypothesis has enjoyed striking success in accounting for and predicting details of photosynthetic and mitochondrial phosphorylation. The greatest test of the imagination in applying this hypothesis has been in concocting mechanisms whereby the pumping of protons through the ATPase drives the conversion of $ADP + P_i$ to $ATP + H_2O$. In this aspect of the problem a hybrid between chemiosmotic and conformational hypotheses may be useful. The binding and release of protons by components of the ATPase might alter the configurations of these components and thus alter their affinities for H_2O in a predominantly water-free (lipoidal) environment. This could control the direction of synthesis or hydrolysis of ATP. Evidence for functional conformational changes in the ATPase will be considered in Section 10.3.

10.2 Light-induced electrochemical gradients in photosynthetic membranes

A clear relationship between light-induced proton uptake by chloroplasts and their capacity for ATP formation was first demonstrated by A. T. Jagendorf, who showed that the two processes have identical kinetics (Fig. 10.3). In this demonstration H^+ uptake was measured with an external pH electrode. The capacity for ATP formation could be assayed by adding $ADP + P_i$ (previously withheld) at any time during the experiment, terminating the illumination if it was on at that time, and measuring the amount of ATP formed subsequently. One standard basis for assaying the formation of ATP is the incorporation of radioactive $^{32}P_i$.[2]

In a more dramatic experiment, Jagendorf and Uribe[3] showed that ATP can be made in response to an artificially imposed pH gradient across the thylakoid membrane, without the help of light. Chloroplasts were soaked in a medium containing succinic acid at pH 4 so as to drive the interiors of the thylakoids to this pH (succinic acid was chosen because in its undissociated form it can permeate the thylakoid membrane). The external pH was then raised abruptly to 8, creating a differential of four pH units across the membrane, and ADP + P_i was added. The resulting ATP formation, independent of light and unaffected by inhibitors of electron transport such as DCMU, was quantitatively spectacular: up to 100 ATP for every Cyt f (hence for every reaction center). This experiment gave great impetus to general interest in Mitchell's chemiosmotic hypothesis.

Many years later Gräber, Schlodder, and Witt[4] showed that the other component of pmf, the membrane potential $\Delta\psi$, can yield quantitatively significant ATP formation when imposed artificially. The artificial electric gradient was created by suspending chloroplasts between electrodes 1 mm apart and applying a potential of 200 V, which was estimated to produce a gradient of about 0.3 V across the thylakoid membrane (Fig. 10.4). The gradient had the correct polarity for ATP formation on one side of the thylakoid; the fact that the polarity was wrong on the opposite side presumably reduced the yield of ATP about twofold.

In dealing with proton uptake and release by thylakoids or chromatophores, the assay of H^+ removed from or returned to the external medium is relatively simple; one need only measure the change of pH with an electrode or, with better time resolution, with a pH-indicating dye (such as phenol red) that does not penetrate the membrane. This method can be calibrated by adding known amounts of strong acid or base. To estimate changes of internal pH is a more delicate matter. One approach is to add a weak base such as methylamine, which can enter the thylakoid in its undissociated form (CH_3NH_2) and can then interact with protons; $CH_3NH_2 + H^+ \longleftrightarrow CH_3NH_3^+$. The principle of this method (see later, Eq. 10.7) also illustrates the action of one category of uncouplers, of which the prototype is ammonia: NH_3 (added as NH_4Cl) can permeate the thylakoid, and on both sides of the membrane it interacts with protons; $NH_3 + H^+ \longleftrightarrow NH_4^+$. A light-induced pH gradient can be washed out because NH_3 can bind H^+ on the inside and NH_4^+ can release H^+ on the outside:

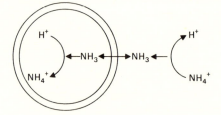

equivalent to a one-for-one exchange of NH_4^+ for H^+:

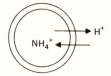

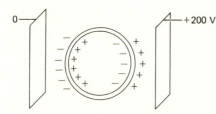

Fig. 10.4. Creation of artificial electric gradient for ATP formation. (See text.)

The system comes to equilibrium when the ratio of $[H^+]$ to $[NH_4^+]$ is the same on each side of the membrane:

$$\frac{[H^+]_i}{[NH_4^+]_i} = \frac{[H^+]_o}{[NH_4^+]_o} \tag{10.5}$$

The final pH difference across the membrane will depend on the amount of NH_4^+ that has been added. The outcome can be predicted by imagining two compartments, the aqueous space inside the thylakoids, of volume V_i, and the exterior space, of volume V_0. The initial interior and exterior proton concentrations are $[H^+]_i$ and $[H^+]_o$; if this initial condition represents a steady state during illumination, $[H^+]_i > [H^+]_o$. NH_4Cl is added to give an initial external concentration $[NH_4^+]_o$, and the exchange of NH_4^+ for H^+ begins. If these initial conditions are applied to Equation 10.5, with the stipulation that the virtual flux of H^+ in one direction equals that of NH_4^+ in the other direction, one can show (left as an exercise for the student) that

$$\Delta pH = \log\left([H^+]_i/[H^+]_o\right) \longrightarrow \log\left([H^+]_i/([H^+]_o + [NH_4^+]_o)\right)$$
$$\text{(initial)} \qquad\qquad\qquad \text{(final)}$$

where the final equilibrium condition is written in terms of the initial concentrations of H^+ and NH_4^+. This result predicts that the difference in pH will be reduced to zero if NH_4^+ is added at a concentration equal to the initial difference in internal and external H^+ concentrations. If the pH is initially 5 inside and 7 outside, NH_4^+ should be an effective uncoupler when added at about 10^{-5} M concentration.

In fact, amines and NH_4^+ act as uncouplers only at concentrations greater than about 10^{-3} M. The interiors of the thylakoids contain such an abundance of H^+ buffers that for every 100 H^+ pumped in by light, about 99 are bound to buffering sites such as $-COO^-$ and $-NH_2$ (the external medium and the exterior surfaces of the thylakoids also have an abundance of buffering sites).[5] The reaction $H^+ + NH_3 \longrightarrow NH_4^+$ inside the thylakoids does not alter the pH greatly because more H^+ is released by the buffering sites:

$$\begin{cases} H^+ + NH_3 \longrightarrow NH_4^+ \\ XH \longrightarrow X^- + H^+ \end{cases}$$

where X^- represents a buffering site. By the time most of these sites have been unloaded, the concentration of NH_4^+ inside the thylakoids has become so high that the consequent osmotic uptake of H_2O threatens to burst the membranes. The membranes then become leaky, allowing NH_4^+ as well as H^+ to move in and out freely. This abolishes the pH gradient and also causes the collapse of any membrane potential that has built up.

If one adds an amount of NH_4^+ or amine well below the concentration that causes uncoupling, the pH is altered very little. The amine becomes distributed according to Equation 10.5 regardless of the presence of internal buffers, and if this distribution can be measured, the difference of pH between inside and outside can be evaluated: Modifying Equation 10.5,

$$\Delta pH = \log([H^+]_i/[H^+]_o) = \log([R-NH_3^+]_i/[R-NH_3^+]_o) \quad (10.6)$$

If n_0 represents the amount of amine added initially to an exterior volume V_0, the initial external concentration $[R-NH_3^+]_o$ is n_0/V_0 (provided that the amount of unprotonated amine, $R-NH_2$, is negligible at the prevailing pH; see later in this paragraph). If x is the amount of amine that moves to the inside during equilibration, the final concentrations are $(n_0 - x)/V_0$ outside and x/V_i inside. Equation 10.6 then becomes

$$\Delta pH = \log\left[\frac{x/V_i}{(n_0 - x)/V_o}\right] = \log\left(\frac{f}{1-f}\right) + \log\left(\frac{V_o}{V_i}\right) \quad (10.7)$$

where f is the fraction of amine, x/n_0, that has been carried into the interiors of the thylakoids. To determine ΔpH under steady-state conditions one must therefore measure both f and V_o/V_i. The volume ratio can be determined by mixing a suspension of chloroplasts with a substance that enters the interior thylakoid space freely. After equilibration the chloroplasts are separated quickly from the external medium, by filtration or centrifugation, and the amount of substance in each phase is determined. The uptake of amine can be evaluated similarly, by separating the chloroplasts from the medium and assaying total amine in each phase. The assay is simplified by using ^{14}C-labeled amine. It can usually be assumed that the amount of unprotonated amine, $R-NH_2$, is negligible, as the pH is well below the pK of a suitable amine such as methylamine.

An exceptionally easy way to measure amine uptake, and one that is well suited to dynamic measurements, is to use a fluorescent amine such as 9-aminoacridine. When this amine enters the thylakoid, its fluorescence is quenched, probably as a result of binding to protein, so the amounts of internal and external amine can be monitored continuously by the changes of fluorescence. There is concern with this method, in that significant amounts of internally bound amine may not participate in the proton-binding equilibrium; this would invalidate Equations 10.5 and 10.6 when based on total amine uptake.

The measurement of internal pH by means of a permeating ind such as neutral red can be useful in some contexts. These ind amines, so they tend to perturb the system that they monitor. useful in rapid kinetic measurements to show proton uptake in response to single flashes. It must be assumed that the indicator has free access to the regions where H^+ is pumped in, and that its spectral response to changes in pH is not altered when it is incorporated into the thylakoid.

An uncoupler whose action is like that of amines is nigericin, a large (about 40 carbons) organic molecule with an affinity for the thylakoid membrane. Nigericin binds both H^+ and K^+, competitively and reversibly. Presumably shuttling back and forth, it can carry H^+ and K^+ in and out of the thylakoid. It cannot transport net charge; each transaction is limited to the substitution of one H^+ for one K^+ or vice versa. By adding nigericin and K^+ to a suspension of chlorplasts, one can promote the replacement of H^+ by K^+ inside the thylakoid. Equilibrium is reached when $[H^+]_i/[K^+]_i = [H^+]_o/[K^+]_o$.

The application of these techniques has shown that if consecutive flashes are applied to chloroplasts, each flash deposits two H^+ inside the thylakoid for every electron transferred from the O_2-evolving system through the PQ pool. One H^+ is released by the O_2-evolving system, and one is moved in by the PQ pool. Owing to experimental uncertainties it is conceivable that this conclusion is wrong, and that the PQ pool translocates two H^+ per electron. In saturating continuous light, chloroplast thylakoids develop an internal pH of about 4.5 against an external pH of 8. The difference of 3.5 pH units amounts to 0.21 V of proton potential. At external pH 8, about 400 H^+ are stored at this potential for every photosynthetic unit (defining such a unit as one P680, one P700, and all the associated chemical machinery). The proton potential is actually a bit more (up to 0.22 V) because the thylakoid sustains a small membrane potential in the light, of no more than 0.01 V. Methods for estimating this potential will now be considered.

Numerous correlations show that changes of membrane potential in chloroplasts and chromatophores are signaled by shifts of the absorption bands of antenna pigments, especially carotenoids. In chloroplasts the main effect is an increase of absorption centered at 518 nm and a decrease at 480 nm, corresponding to the red shift of an absorption band of β-carotene. In purple bacteria the spectra of the band shifts vary according to the type of carotenoid present, and there is evidence (see later in this section) that only a specific minor component of the total carotenoid engages in the shift. Chls *a* and *b* and Bchl also show these shifts, but the carotenoid shifts have been the easiest to measure without interference from other optical changes and have been studied and exploited far more than the shifts of chlorophylls.

Reversible light-induced shifts of carotenoid absorption bands were discovered[6] long before their origin was understood, but it was natural to speculate that they were induced by electric fields related to charge separation in the membrane. This interpretation was consistent with the very rapid onset of the carotenoid shift at the start of illumination; at least a part of the shift could be associated with the primary charge separation. It was also recognized that the carotenoid shift in chromatophores of purple bacteria is lost when the physical integrity of the membrane is attacked, as by the addition of a detergent. Correlations then emerged between the carotenoid shift and the factors that affect ATP formation. The shift is abolished (more correctly, its decay after illumination is accelerated) by some types of uncoupler, and it is affected by substrates and inhibitors of ATPase. These observations pointed toward an association between the carotenoid shift and the "membrane potential" component, $\Delta\psi$, of the proton potential. Most convincingly, a carotenoid shift having the same spectrum as the light-induced shift can be induced in chromatophores by chemical treatments designed to create an ionic gradient (and hence an electric field) across the membrane, and the shift is eliminated by conditions that should nullify the gradient.

This brings us to another category of uncoupler, typified by valinomycin, a large organic molecule shaped like a ring with a hydrophilic interior. Valinomycin can make artificial lipid membranes selectively permeable to specific cations, including K^+ and NH_4^+ but not Na^+ or H^+, and it appears to have the same effect on photosynthetic membranes. One can imagine that a molecule of valinomycin in the membrane creates a hole for the free passage of the permitted ions. If chromatophores are exposed to valinomycin and a high external concentration of KCl, the selective influx of K^+ creates a charge imbalance (a diffusion potential) with an excess of K^+ inside and a corresponding excess of Cl^- outside:

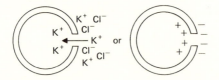

Equilibrium is reached when the electric force, resisting the influx of more K^+, balances the difference in K^+ concentration. The equilibrium is expressed by a form of the Nernst equation,

$$\Delta\psi \text{ (volt)} = 0.06 \log \left([K^+]_o / [K^+]_i \right) \qquad (10.8)$$

This treatment causes a carotenoid shift of magnitude proportional to $\Delta\psi$ as calculated from Equation 10.8. Jackson and Crofts[7] used this technique to calibrate the carotenoid shift in terms of membrane potential, by adding pulses to KCl to chromatophores of *Rp. sphaeroides* treated with valinomycin and observing the size of the consequent absorbance change (Fig. 10.5). The response was proportional to the computed $\Delta\psi$ over a thousandfold range of $[K^+]$ applied externally.

Using this calibration, it appears that saturating continuous light causes *Rp.*

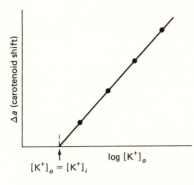

Fig. 10.5. Carotenoid shift as a function of KCl added to chromato-phores of *Rp. sphaeroides* in the presence of valinomycin.

sphaeroides chromatophores to develop a membrane potential of about 0.2 V in the steady state, after an initial surge to about 0.4 V. A single-turnover flash induces a carotenoid shift corresponding to about 0.1 V. The earliest onset of this shift should be attributed to the primary charge separation (this can be seen even at liquid-helium temperature), but within a few milliseconds this is transmuted into a displacement of H^+ across the membrane. The pH differential in illuminated chromatophores is about one unit, so the total pmf is about 0.25 V, close to that in chloroplasts, but the greater part of this proton potential resides in the membrane-potential component.

Calibration of the carotenoid shift in chloroplasts by means of valinomycin and KCl has proved more difficult than in chromatophores, because of inter-ference by optical effects (especially changes of light scattering, of osmotic origin) and by the movements of permeating ions such as Cl^-. The results of such calibrations first indicated that the translocation of one H^+ per photo-synthetic unit gives a $\Delta\psi$ of about 15–35 mV, and the steady-state $\Delta\psi$ in continuous light is less than 0.1 V. These values have been revised downward; $\Delta\psi$ (steady state) < 0.01 V on the basis of ion-distribution measurements. Some of the carotenoid shift should probably be attributed to potential gradients parallel to the thylakoid membrane and not to a transmembrane potential.

The relative contributions of ΔpH and $\Delta\psi$ to the light-generated pmf in chloroplasts and chromatophores are indicated by the effects of the two cate-gories of uncouplers: those that eradicate ΔpH, exemplified by nigericin (which exchanges K^+ for H^+), and those such as valinomycin that attenuate $\Delta\psi$ by allowing certain cations to diffuse passively through the membrane. The combination of both types can be especially effective in uncoupling elec-tron transport from phosphorylation, as can a third class of uncoupler that dissipates both components of pmf by promoting the passive diffusion of H^+.

This group includes detergents, osmotic shock, and chemicals such as gramicidin (a polypeptide) and carbonylcyanide m-chlorophenylhydrazone (CCCP). The prevailing view of the difference between chloroplasts and chromatophores, in the balance between ΔpH and $\Delta\psi$, has been that the thylakoid membrane is more permeable than the chromatophore membrane to ions such as Cl^- and Mg^{2+}, or that a greater concentration of these ions is available in and around the thylakoids. However, the relative sizes of these vesicles may also play a part. For a given shape, a smaller vesicle has a higher surface:volume ratio than a larger one (for a sphere, $S/V = 3/R$). The translocation of protons is proportional to the surface, and so is the buffer capacity if the buffering sites are on the interior surface. The capacity for ions such as Cl^- and Mg^{2+} is proportional to the volume, so the extent to which such ions can neutralize an imbalance of charge diminishes as the vesicle becomes smaller. There is some evidence for this; well-treated "intact" chloroplasts show a higher proportion of ΔpH relative to $\Delta\psi$ than smaller thylakoid fragments, and chromatophores are much smaller than chloroplast thylakoids. If the surface:volume ratio is important in this way, the relative contribution of ΔpH to the pmf should be greater in the intact living tissue than in the vesicular membrane fragments with which the storage of proton potential is usually measured.

There appear to be no significant barriers to the diffusion of H^+ inside a thylakoid or chromatophore. An uncoupler such as gramicidin, applied at a concentration equivalent to one molecule per chromatophore or thylakoid, can have a pronounced uncoupling effect, as if a single hole in the membrane can let most of the extra internal H^+ leak out. A typical chromatophore has about 15 to 50 reaction centers (about 1000 to 5000 Bchl molecules), and a typical thylakoid has about 300 photosynthetic units (about 10^5 chlorophylls).

Reports of a "critical threshold" value of pmf, below which phosphorylation cannot occur, are questionable. The impression of the threshold may have been engendered by a nonlinear relation between the rate of ATP formation and the concentration of protons available at the ATPase. If three H^+ must be moved out through the ATPase for every ATP formed, one can expect the rate to vary as $[H^+]^3$. The shape of a plot of this cubic dependence could suggest a "critical threshold" concentration of H^+, below which ATP is not formed.

The imputed effect of membrane potential in causing shifts of the absorption bands of carotenoids is usually visualized as a molecular counterpart of the classical Stark effect in which the ground- and excited-state energy levels of atoms are shifted by an external field. In atomic spectra the frequencies of absorption and emission are shifted in proportion to the square of the field strength; $\Delta\nu \propto E^2$. This can be explained qualitatively by saying that the field acts in two ways. The field first induces electric dipoles (or orients permanent dipoles in the direction of the field), with the dipole moment proportional to

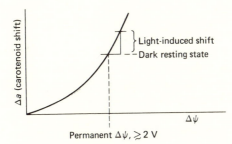

Fig. 10.6. Plot of Δa vs. $\Delta\psi$ showing pseudolinearity of the light-induced shift. (See text for discussion.)

E, and the shift in energy levels is then proportional to the product of dipole moment and field strength. The frequency is altered if the dipole moment in the ground state is different from that in the excited state, so that the energies of these states are changed to different extents. In contrast, the band shifts of carotenoids appear to be linear rather than quadratic functions of the field strength (of $\Delta\psi$). One can rationalize this by saying that the carotenoids have permanent dipole moments, larger than any induced moments, and fixed orientations in the membrane. The first of the two sources of dependence on field strength is then eliminated. However, the nearly symmetric structures of carotenoids that show band shifts in vivo argue against the presence of large permanent dipole moments. Alternatively, one can say that the dependence is quadratic, but a strong permanent field across the membrane places the observed variations on a nearly linear part of the curve (Fig. 10.6). The existence of at least a small permanent field is indicated by the fact that some chemical treatments of chromatophores, designed to produce ionic diffusion potentials of "reverse" polarity (negative inside), cause blue-shifts rather than red-shifts of the carotenoids. However, this explanation of a pseudolinear relation between the carotenoid shift and the membrane potential would require a permanent membrane potential of the order of 2 V.

Another puzzle is the complete absence of carotenoid shifts in organisms such as *Rp. viridis* that ought to exhibit them. Possible explanations based on "wrong type of carotenoid" or "wrong orientation" seem unlikely, but have not been eliminated rigorously. It may be that the carotenoid shift is caused by factors other than, but related to, the alleged membrane potential. These could include a slight displacement of the carotenoid to a medium of greater dielectric constant, or a slight distortion of the molecule around the usual *cis-trans* loci. It is probably simplistic to visualize the membrane as one would imagine a charged parallel-plate condenser, with a uniform field perpendicular to the surface.

In *Rp. sphaeroides* the light-induced band shifts of carotenoids amount at most to about 1 nm, and are exhibited only by a special component, the "shifting pool," comprising about 15% of the total carotenoid. The "shifting" carotenoids are chemically the same as the nonshifting ones, but their absorption maxima are at wavelengths about 7 nm greater than those of their non-shifting counterparts. This again suggests the presence of a strong permanent field, sufficient to cause a 7-nm shift (in comparison with a maximum light-induced shift of only 1 nm). The amount of shifting pool is correlated with the amount of the B850 antenna component; the carotenoids associated with B875 do not seem to shift. This is especially clear in mutants of *Rp. capsulata* that lack either B850 or B875.

We have yet to consider the delayed fluorescence of Chl or Bchl as an indicator of proton potential in photosynthetic tissues. Delayed fluorescence of Chl in algae was discovered by Strehler and Arnold[8] when they attempted to use the "firefly luminescence" assay to detect light-induced ATP formation in *Chlorella*. They placed the algae in a syringe, illuminated them (or, as a control, kept them in the dark), and then injected them rapidly into a cell containing firefly extract. The cell was in a dark box viewed by a light detector. The mixture in the cell emitted light when the algae were injected, but only if they had first been illuminated. That the light came from Chl in the algae and not from luciferin was discovered when firefly extract was (inadvertently?) not put into the cell prior to the injection of the algae.

Delayed fluorescence in photosynthetic tissues results from recombination of P^+ and a reduced early (primary or secondary) electron acceptor in the reaction center, producing P^* (with the help of some energy from the surroundings). This is followed by migration of the quantum out into the antenna, and finally emission of fluorescence from the resultant Bchl* in the antenna. We are confident of this interpretation for three reasons. First, the intensity of the delayed fluorescence can be correlated with the concentrations of the precursors P^+ and reduced acceptor. Second, the yield of delayed fluorescence from isolated reaction centers is extremely low; a quantum must have an opportunity to become "lost" in an antenna in order for delayed fluorescence to be observed easily in photosynthetic preparations. Third, there is a sound precedent in the behavior of purified Bchl dissolved in methanol. Bchl in vitro emits a luminescence having the same spectrum as its fluorescence when it is reduced by Fe^{2+} after having been oxidized by Fe^{3+}:

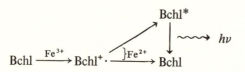

The yield of this luminescence is low, about 10^{-3}, because the reduction of Bchl$^+$· usually produces Bchl in its ground state. Only rarely does the reduced product appear in its excited state, Bchl*.

The intensity of delayed fluorescence is influenced by two major factors aside from the concentrations of precursors. Once a quantum has escaped from P* in the reaction center and entered the antenna, the likelihood that it will emerge as fluorescence is governed by the same factors (see Section 2.5) that control the yield of primary fluorescence. A qualification is that the quantum is close to an "open" reaction center, the reaction center that had produced it. It is therefore more likely to be trapped into the photochemical route than a quantum absorbed at random in the antenna. More important for our present considerations, the intensity of delayed fluorescence is affected strongly by both components of the proton potential.

The effect of a membrane potential (positive inside, as induced by light) is to intensify the delayed fluorescence. The intensification is an exponential function of $\Delta\psi$. This has been established by measuring both the carotenoid shift and the delayed fluorescence under a variety of conditions (including the use of uncouplers) that alter $\Delta\psi$. An interpretation of the exponential dependence was offered by Crofts et al.[9] as follows. It is presented for the bacterial couple P^+, Q_a^-, but it applies in the same way for any combination of P^+ and reduced acceptor that serves as a source of delayed fluorescence. In the absence of a membrane potential, the gap in energy between P^+, Q_a^- and P^*, Q_a can be written ΔE. The rate of conversion from P^+, Q_a^- to P^* is limited by a Boltzmann factor, $e^{-\Delta E/kT}$. When a membrane potential $\Delta\psi$ is present, the energy of P^+, Q_a^- is greater. An electron, in moving from P to Q_a, is raised through a difference of potential $\Delta\psi$, and the corresponding energy (equal to $\Delta\psi$ in electron volts) is stored in P^+, Q_a^-, over and above the energy that is stored in the absence of a membrane potential. The energy gap between P^+, Q_a^- and P^*, Q_a is correspondingly less; it is $\Delta E - \Delta\psi$:

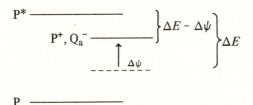

The Boltzmann factor becomes $e^{-(\Delta E - \Delta\psi)/kT}$ or $(e^{-\Delta E/kT})(e^{\Delta\psi/kt})$; this factor governs the rate of formation of P* and hence the intensity of delayed fluorescence. Thus the delayed fluorescence is intensified by a factor $e^{\Delta\psi/kT}$ when a membrane potential $\Delta\psi$ is present.[10]

One might expect that the extra energy derived from $\Delta\psi$ and stored in P^+, Q_a^- can be utilized, but there is no clear evidence for this. The extra energy might be lost as the electron on Q_a^- continues through its cycle.

The other component of proton potential, ΔpH, also intensifies the delayed fluorescence. The reason for this might be that the concentration of

H^+ affects the equilibria of reactions that in turn alter the concentration of Q_a^-, or of any other reduced acceptor that serves as a substrate for delayed fluorescence. The effects of $\Delta\psi$ and ΔpH on the delayed fluorescence of chloroplasts have been sorted out by the selective use of nigericin and valino-mycin as uncouplers.[11] Recent measurements of delayed fluorescence from chloroplasts in response to single flashes, with chemical treatment to induce a membrane potential and/or a pH gradient, have introduced the complication that the ability of a membrane potential to enhance the delayed fluorescence requires the simultaneous presence of a pH gradient or perhaps a low pH inside the thylakoids.

The measurement of delayed fluorescence has been an interesting adjunct to the study of the two components of proton potential, but the delayed fluorescence responds so promiscuously to a multitude of factors that it has played a supportive rather than a primary role in helping us to understand energy coupling in photosynthesis.

In reviewing our knowledge of electrochemical gradients in photosynthetic tissues, much of it seems to be based on crass empiricism, and yet we can be quite confident that proton potential is the energetic link between electron transport and ATP formation. Present experimental evidence, interpreted through the chemiosmotic hypothesis, provides a very satisfactory description of the coupling of electron flow and phosphorylation on the basis of proton fluxes.

10.3   ATP formation and the energy efficiency of photosynthesis

The enzyme system that forms ATP is called the ATPase complex or simply ATPase. In chloroplasts this complex protein consists of a water-soluble part called CF_1 (for chloroplast coupling factor one) that is readily removed from the thylakoid surface, and a hydrophobic part F_0 that is embedded in the membrane and appears to be a channel for H^+ as well as a binding site for CF_1. This terminology was adapted from the complex F_1-F_0 that mediates respiration-linked or oxidative phosphorylation in mitochondria. Bacteria have an ATPase complex similar to the F_1-F_0 of mitochondria, and in photosynthetic bacteria the F_1-F_0 complexes that mediate oxidative and photosynthetic phosphorylation have thus far proved to be very simi-lar. In thylakoids CF_1 makes up about a tenth of the membrane protein; there are about three of these entities for every photosynthetic unit.

We shall consider the H^+/ATP stoichiometries and energy balance of these proton-translocating, ATP-forming enzyme systems and then some aspects of their composition, structure, and mechanism of action. We have more specu-lation than factual knowledge about the mechanism.

The ratio of H^+ translocated to ATP formed has been estimated by measur-ing the net proton efflux (out of the thylakoids, through the ATPase) that can be correlated with ATP synthesis. This has been done with chloroplasts by establishing a basal rate of light-driven H^+ uptake, in the absence of added ADP and hence with minimal synthesis of ATP, and then seeing to what extent the net H^+ uptake is diminished when ADP (and P_i) is added. The ADP-induced reduction of net H^+ uptake is equated to the efflux linked to ATP formation. The amount of ATP formed in response to the added ADP is also measured, and the ratio of H^+ exported to ATP synthesized can thus be found.

Another approach is to monitor the decay of a flash-induced carotenoid band shift, and to measure the extent to which this decay is accelerated by added ADP + P_i. If the band shift has been calibrated in terms of the net difference of charge (i.e., of H^+) across the membrane, the ADP-induced export of H^+ can be computed. It must be assumed that the added ADP + P_i does not alter the rate at which membrane potential (the correlative of carotenoid band shift) is converted to ΔpH by passive ion diffusion. This approach is especially applicable to chromatophores, for reasons given in the last section.

The results of these measurements with chloroplasts[12] are converging to a value of $3H^+$/ATP, in contrast to the $2H^+$/ATP found for oxidative phosphorylation in mitochondria. Support for the ratio $3H^+$/ATP was found by Portis and McCarty,[13] who showed that the rate of synthesis of ATP in chloroplasts varies as the third power of the concentration of H^+ that drives the reaction. In chromatophores of *Rp. sphaeroides* and *Rp. capsulata* the ratio was reported[14] to be $2H^+$/ATP, as for oxidative phosphorylation, but this value has not been established beyond question. These H^+/ATP ratios enter into the energy budget of phosphorylation because the source of chemical free energy for ATP formation is the flow of H^+ down the gradient of proton potential across the membrane.

The free energy stored as ATP, and available when ATP is hydrolyzed to ADP + P_i, is given by

$$-\Delta G' \text{ (kcal mole}^{-1}) = -\Delta G_0' + 1.36 \log \left(\frac{[\text{ATP}]}{[\text{ADP}][\text{P}_i]} \right) \qquad (10.9)$$

In this equation $\Delta G_0'$ is the standard free energy (see any proper text on chemical thermodynamics) of hydrolysis of ATP. It is negative (and $-\Delta G_0'$ is positive) because the hydrolysis of ATP releases energy, and by convention $\Delta G_0'$ is the energy stored in the reaction. "Standard," denoted by the subscript 0, means that all reactants and products are at unit activity,[15] one molar. The actual $\Delta G'$ for the prevailing activities of ATP, ADP, and P_i is computed from the standard $\Delta G_0'$ by Equation 10.9. The "prime" superscript means that $\Delta G'$ and $\Delta G_0'$ are specified under laboratory conditions different from the formal standard conditions, but in ways that do not invalidate Equation 10.9. If this equation is written in units of electron volts per molecule rather than kilocalories per mole, the numerical factor 1.36 should be replaced by 0.06. A currently accepted value of $\Delta G_0'$ for ATP hydrolysis is -7.8 kcal mole^{-1} or -0.34 eV molecule^{-1}.

Measurements of the concentrations of ATP, ADP, and P_i in chloroplasts exposed to saturating illumination under optimal conditions have given a value of $-\Delta G'$, computed from Equation 10.9, of about 14 kcal mole^{-1} or 0.6 eV molecule^{-1}. This is the maximum free energy expended, by H^+ moving out of the thylakoids, in forming one molecule of ATP. If three H^+ are exported per ATP, each H^+ must move through a difference of 0.6/3 or 0.2 V of proton potential. This conclusion is compatible with the pmf estimated from ΔpH (3.5 units, equivalent to 0.21 V) and $\Delta\psi$ (<0.1 V) as described in the last section.

The value of $-\Delta G'$ for photosynthetic ATP formation in chromatophores of *Rhodospirillum rubrum* has been estimated at 12.1 to 12.4 kcal mole^{-1} (0.53 V), and the pmf has been found to be about 0.2 V through measurement of ^{14}C-amine uptake driven by the proton gradient. Two H^+ would yield 0.4 V, less than the required 0.53 V. Perhaps the formation of one ATP requires the translocation of $3H^+$ rather than $2H^+$.

The mechanism of proton-driven ATP formation remains vague and speculative; both "chemical" and "conformational" hypotheses have been advanced. Detailed chemical schemes, as proposed for the past 20 years, have gone through numerous variations. One such sequence, with ADP written $R—O^-$ as in Reaction 10.2, is shown in Figure 10.7. This reaction scheme fits naturally with a requirement of $2H^+/ATP$; it does not show the role of the third H^+ that seems to be required in phosphorylation by chloroplasts. Studies with the stable isotope ^{18}O have shown that the oxygen atom bridging R and P, in the ATP as represented here, comes from ATP and not from P_i.

The most popular current form of a conformational hypothesis holds that the ATP is made in a reaction between enzyme-bound ADP and P_i that requires little or no energy. The major energy-requiring step is the release of the newly formed ATP, caused by a change in the configuration of the enzyme (Fig. 10.8). There is evidence (see later in this section) that the CF_1 of chloroplast ATPase does change its shape reversibly when illuminated, but the change could be incidental rather than fundamental to the mechanism of ATP formation.

In studying any enzyme one hopes that the mechanism of its action will become clear when its composition and structure have been fully elucidated. At this writing we have learned much about the compositions of the ATPase complexes CF_1-F_0 and F_1-F_0, a little about their structures, and almost nothing definitive about how they fulfill their function.

Thanks to the pioneering efforts of E. Racker and collaborators, the various protein components of F_1-F_0 and CF_1-F_0 can be isolated, separated, and characterized. We are approaching the ability to reconstitute the separated polypeptides into a functioning ATPase complex (as a single example see Kagawa and Racker in the suggested reading). CF_1 is washed away from chloroplast thylakoids by a solution of the cation-binding agent EDTA, and can be replaced with the help of Mg^{2+}. Removal of F_1 from chromatophores

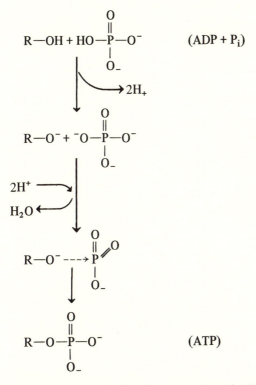

Fig. 10.7. A scheme for the conversion of ADP + P_i to ATP + H_2O, with $2H^+$ released and bound in successive steps.

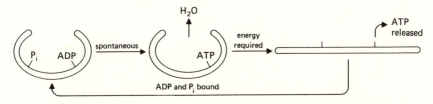

Fig. 10.8. A scheme for the reaction of enzyme-bound ADP and P_i to form ATP, with release of ATP dependent on a change in conformation of the enzyme.

of *Rp. capsulata* and *Rs. rubrum* requires sonic disruption in the presence of EDTA, but chromatophores of *Chr. vinosum* release their F_1 when simply washed with a buffer solution. CF_1 is bound to the thylakoid membrane by a hydrophobic protein F_0, an integral part of the membrane. F_0 can be removed from the membrane by means of detergents, and the isolated F_0 will recombine spontaneously with isolated CF_1 to form a CF_1-F_0 complex. The

CF_1-F_0 complex can also be isolated from the membrane by the use of detergents. This complex is functional; its ATP-forming ability can be demonstrated in artificial phospholipid vesicles in which the complex has been embedded. To show this one must endow the vesicles with proton pumps: either fragments of membrane active in respiratory electron transport or bits of the purple membrane of *Halobacterium halobium*, which pump protons in response to light (see Section 6.4).

The removal of CF_1, leaving F_0 unmasked, makes the thylakoid membrane permeable specifically to H^+ ions. This implies that F_0 is a channel or a carrier to conduct protons from inside the thylakoid to CF_1. F_0 is made up of three distinct polypeptides of apparent weights (by SDS–polyacrylamide gel electrophoresis) 8, 13, and 19 kdalton. It shows no ATP/ADP reactivity in its isolated form. The 8-kdalton polypeptide is probably responsible for H^+ permeation; it interacts with agents that prevent H^+ permeation.

Isolated CF_1, a water-soluble protein weighing 325 kdalton, can be made to display ATPase (ATP-hydrolyzing) activity. The activity must be induced, either by heat or by reaction with sulfhydryl reagents (chemicals with free —SH groups), and is maintained only if Mg^{2+} or Ca^{2+} is present. This might be related to the fact that the "true" substrate of ATP hydrolysis is the ATP-Mg^{2+} complex. The ATPase activity has been of great help in the study of isolated CF_1 and its components. The ATPase activity of CF_1 in its native setting will be considered later in this section, in connection with conformational changes.

CF_1 can be separated into its component polypeptides by exposure to urea or guanidine, agents that disrupt hydrogen bonds (weak electrostatic bonds in which an H atom is a bridge). It helps to use a detergent as well, and the detergent SDS is effective by itself. SDS–polyacrylamide gel electrophoresis shows five distinct polypeptides of apparent weights ranging from 14 to 60 kdalton. These polypeptides have been characterized as to their amino acid content and probable number per CF_1, and their individual functions have been inferred, although not proved, by methods to be described. The names, weights, proportions (tentative), and presumed functions of these subunits of CF_1 can be tabulated (Table 10.1). The content of Table 10.1 was culled from one presented by McCarty in an excellent recent review (R. E. McCarty, 1979, suggested reading).

The functions of the CF_1 subunits have been explored by examining the activities of partly degraded CF_1 (bereft of some subunits) and partly reconstituted groups of subunits. For example, tryptic digestion of CF_1 yields a preparation of almost pure $\alpha\beta$ complexes, and these have ATPase activity. The effects on CF_1 of antibodies specific to individual subunits have been examined, and also the interactions among subunits and between subunits and chemicals that affect ATPase activity. Details of these studies, which have suggested the functions listed in the table, can be traced through McCarty's 1979 review.

Table 10.1. *Properties of* CF_1 *subunits*

Subunit	Apparent weight (kdalton)	Probable number per CF_1	Provisional assignment of function
α	60	2	Regulatory
β	56	2	Active site
γ	37	1	Carries H^+ to β
δ	19	1	Attachment to membrane
ϵ	14	2	Inhibitor of reverse reaction; attachment to membrane

Light causes a reversible conformational change of CF_1 in chloroplasts, as shown most strikingly by the experiments of Ryrie and Jagendorf[16] on the light-dependent incorporation of tritium, presented as ^3H-labeled water, into the CF_1. Illumination could cause the incorporation, by exchange, of up to 100 atoms of ^3H per CF_1, and these atoms were retained in subsequent darkness. Incorporation of ^3H from water into CF_1 could also be induced by an artificial pH gradient (more acid inside), and uncouplers prevent both the light-induced and the ΔpH-induced incorporation. The same dependence on light, pH gradient, and uncouplers is shown by a reaction of CF_1 with N-ethylmaleimide (NEM), a reagent that reacts primarily with $-SH$ groups. Light (or ΔpH) exposes $-SH$ groups that are not accessible to the NEM in the dark. Thus both ^3H incorporation and the exposure of $-SH$ groups are induced in CF_1 when the thylakoid develops a proton potential. The conclusion drawn from binding of NEM must be qualified by the fact that NEM reacts, although more slowly, with ϵ-amino groups of lysine in the protein.

A conformational change is also suggested by the induction of ATP-hydrolyzing activity in chloroplasts. Normally the CF_1 in vivo has very little of this ATPase activity, even when the reversal of ATP synthesis should be favored thermodynamically (at high ATP concentration). The ATPase activity can be induced by illumination of chloroplasts in the presence of sulfhydryl reagents, perhaps because newly exposed disulfide bridges are reduced to $-SH$; $-S-S- \longrightarrow -SH + HS-$. In this reaction, as in the foregoing ones, the action of light can be replaced by ΔpH and is inhibited by uncouplers. The ATPase activity is retained in the dark if Mg^{2+} is present. In these activated chloroplasts the hydrolysis of ATP is attended by inward pumping of H^+ and is *accelerated* by uncouplers; thus the reaction appears to be an authetic reversal of the natural ATP synthesis.

Despite the growing evidence for conformational changes of CF_1 associated with the activity of the ATPase complex, we do not yet have a way to decide whether these changes are essential or merely reflexive to the synthesis of ATP. A more intimate knowledge of the structure of CF_1 might help to unravel this problem. CF_1 and F_1 have now been crystallized, and we can hope that the structures of these macromolecules will eventually be determined through X-ray diffraction analysis. For the present the chemical and physical details of photosynthetic ATP formation seem almost as obscure as those of oxygen evolution.

Various types of ATPase are of central importance in all of life. Many act as ion pumps, using the energy of ATP hydrolysis to exchange H^+ for Na^+, K^+, or Ca^{2+}, or Na^+ for K^+. An ATPase exchanging Na^+ for K^+ against the concentration gradients of these ions enables sensory cells and nerves to function by pumping K^+ into the cells and Na^+ out; an ATPase that pumps Ca^{2+} is essential to the operation of muscle and other contractile proteins. In none of these examples is the mechanism of ion translocation by the ATPase known.

At this point we are in a position to compute the energy efficiency of photosynthesis, starting with the absorbed quanta and ending with the products ATP and NADPH.

Plants

Input: Eight quanta near 680 nm; energy = 8 × 1.82 = *14.6 eV.*
Output: (1) Four electrons from H_2O to NADPH, spanning 1.15 V redox potential. Energy = 4 × 1.15 = *4.6 eV.* (2) Eight H^+ giving 8/3 ATP[17] at 0.34 eV per ATP. Energy (8/3) × 0.34 = *0.9 eV.* Total output energy = *5.5 eV.*
Efficiency = 5.5/14.6 = *38%.*

Purple bacteria (cyclic only)

Input: One quantum at 900 nm (thermally relaxed level of 870 nm absorption band), *1.38 eV.*
Output: If $2H^+$ per electron going through cycle,

$$1e^- \longrightarrow 2H^+ \longrightarrow 1 \text{ ATP}, \textit{0.34 eV.}$$

Efficiency = 0.34/1.38 = *25%.*

Apparently the plants do better by using noncyclic electron transport.

SUGGESTED READINGS

Jagendorf, A. T. (1977). Photophosphorylation. In *Encyclopedia of Plant Physiology*, New Series, Vol. 5, A. Trebst and M. Avron, eds., pp. 307–37. Springer-Verlag, New York.

Chance, B., and Williams, G. R. (1956). The respiratory chain and oxidative phosphorylation. *Adv. Enzymol. 17*, 65–134.

Boyer, P. D. (1974). Conformational coupling in biological energy transductions. In *Dynamics of Energy-Transducing Membranes*, L. Ernster, R. Estabrook, and E. C. Slater, eds., pp. 289–301. Elsevier, Amsterdam.

Mitchell, P. (1961). Coupling of phosphorylation to electron and hydrogen transfer by a chemi-osmotic type of mechanism. *Nature (London) 191*, 144–8.

Greville, G. D. (1969). A scrutiny of Mitchell's chemiosmotic hypothesis of respiratory chain and photosynthetic phosphorylation. *Curr. Top. Bioenerg. 3*, 1–78.

Racker, E. (1973). *Mechanisms in Bioenergetics*. Academic Press, New York. 259 pp.

Witt, H. T. (1979). Energy conservation in the functional membrane of photosynthesis. Analysis by light pulse and electric pulse methods. The central role of the electric field. *Biochim. Biophys. Acta 505*, 355–427.

Avron, M. (1978). Energy transduction in photophosphorylation. *FEBS Lett. 96*, 225–32.

Jackson, J. B., and Dutton, P. L. (1973). The kinetic and redox potentiometric resolution of the carotenoid shifts in *Rhodopseudomonas sphaeroides* chromatophores: Their relationship to electric field alterations in electron transport and energy coupling. *Biochim. Biophys. Acta 325*, 102–13.

Saphon, S., Jackson, J. B., and Witt, H. T. (1975). Electric potential changes, H^+ translocation and phosphorylation induced by short flash excitation in *Rhodopseudomonas sphaeroides* chromatophores. *Biochim. Biophys. Acta 408*, 67–82.

Saphon, S., Jackson, J. B., Lerbs, V., and Witt, H. T. (1975). The functional unit of electrical events and phosphorylation in chromatophores from *Rhodopseudomonas sphaeroides*. *Biochim. Biophys. Acta 408*, 58–66.

Goedheer, J. C., and Vegt, G. R. (1962). Chemiluminescence of chlorophyll *a* and bacteriochlorophyll in relation to redox reactions. *Nature (London) 193*, 875–6.

Hinkle, P. C., and McCarty, R. E. (1978). How cells make ATP. *Sci. Am. 238*, 104–23.

McCarty, R. E. (1979). Roles of a coupling factor for photophosphorylation in chloroplasts. *Annu. Rev. Plant Physiol. 30*, 79–104.

Rosing, J., and Slater, E. C. (1972). The value of ΔG^0 for the hydrolysis of ATP. *Biochim. Biophys. Acta 267*, 275–90.

Kraayenhof, R. (1969). "State 3–state 4 transition" and phosphate potential in "Class 1" spinach chloroplasts. *Biochim. Biophys. Acta 180*, 213–15.

Slater, E. C. (1977). Mechanism of oxidative phosphorylation. *Annu. Rev. Biochem. 46*, 1015–26.

Kagawa, Y., and Racker, E. (1971). Reconstitution of vesicles catalyzing $^{32}P_i$-adenosine triphosphate exchange. *J. Biol. Chem. 246*, 5477–87.

Avron, M. (1963). A coupling factor in photophosphorylation. *Biochim. Biophys. Acta 77*, 699–702.

O'Keefe, D. P., and Dilley, R. A. (1977). The effect of chloroplast coupling factor removal on thylakoid membrane permeability. *Biochim. Biophys. Acta 461*, 48–60.

McCarty, R. E., and Racker, E. (1966). Effect of a coupling factor and its antiserum on photophosphorylation and hydrogen ion transport. *Brookhaven Symp. Biol. 19*, 202–14.

11 Carbon assimilation by plants

11.1   The reductive pentose cycle; photorespiration

The central topic of this section, the reductive pentose cycle, is the main pathway of CO_2 assimilation by plants. It is a major system for incorporation of CO_2 by photosynthetic bacteria (including the O_2-evolving cyanobacteria) and is active in a great variety of microbes, indicating that it evolved in organisms far more primitive than eukaryotic algae and plants. We shall examine the chemical nature of this remarkably intricate cycle, and the methods by which it was elucidated.

The basic method of exploring pathways of CO_2 assimilation was to expose illuminated algae to ^{14}C-labeled CO_2, then kill the algae by plunging them into boiling alcohol (within a few seconds after the [14]CO_2 had been added, or later), then extract and separate as many likely organic compounds as possible, and thus determine the flow of ^{14}C into these compounds as a function of time. An invaluable analytical procedure was two-dimensional chromatography in paper, followed by radioautography of the compounds distributed in the paper. The distribution of ^{14}C among the carbon atoms of a specific compound could also be determined. The [14]CO_2 could be added prior to, during, or after illumination. Alternatively, the compounds involved in CO_2 metabolism could be labeled as uniformly as possible by several minutes of illumination with [14]CO_2, simply to provide a convenient quantitative assay, and changes in their amounts could then be monitored while light and CO_2 concentration were altered as functions of time.

These experiments, made principally with the green algae *Chlorella* and *Scenedesmus*, culminated in the description[1] by M. Calvin and collaborators of the reductive pentose cycle in essentially the form that is accepted today. An abbreviated outline of the cycle is shown in Fig. 11.1. CO_2 is incorporated in a reaction with the phosphorylated sugar ribulose diphosphate (RuDP), giving two molecules of 3-phosphoglyceric acid (PGA):

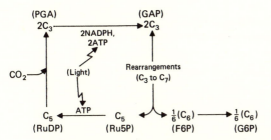

Fig. 11.1. A skeletal outline of the reductive pentose or Calvin cycle. PGA is 3-phosphoglyceric acid; GAP is glyceraldehyde-3-phosphate. F6P and G6P are fructose-6-phosphate and glucose-6-phosphate. Ru5P and RuDP are the mono- and di- (1,5) phosphates of the 5-carbon sugar ribulose. Details of the rearrangements involving 3- to 7-carbon compounds can be found in the suggested reading (Bassham and Calvin, 1957, or Bassham, 1965). Other details are in the text.

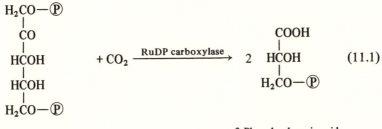

$$\text{Ribulose-1,5-diphosphate} + CO_2 \xrightarrow{\text{RuDP carboxylase}} 2 \;\text{3-Phosphoglyceric acid} \tag{11.1}$$

where ℗ stands for $-\overset{\overset{\textstyle O}{\|}}{\underset{\underset{\textstyle OH}{|}}{P}}-OH$. This reaction is catalyzed by the enzyme

RuDP carboxylase. Each of the two molecules of PGA is then reduced to glyceraldehyde-3-phosphate (GAP), at the expense (for both molecules) of two NADPH and two ATP:

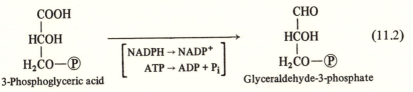

$$\text{3-Phosphoglyceric acid} \xrightarrow[\;ATP \to ADP + P_i\;]{\;NADPH \to NADP^+\;} \text{Glyceraldehyde-3-phosphate} \tag{11.2}$$

This is the only reductive step in the cycle. The GAP enters into an intricate set of rearrangements involving C_3 to C_7 compounds; see the legend of

Fig. 11.1 for references. As a result the six carbon atoms of two GAP are redistributed so that one ends up in fructose-6-phosphate and the other five emerge as rubulose-5-phosphate (Ru5P). The phosphorylated fructose is converted to glucose-6-phosphate in a peripheral reaction, and the Ru5P is converted to RuDP at the expense of one more ATP:

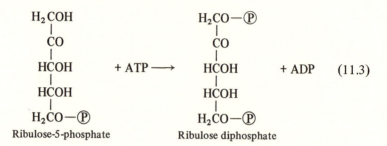

$$\begin{array}{ccc}
\text{H}_2\text{COH} & & \text{H}_2\text{CO}-\text{\textcircled{P}} \\
| & & | \\
\text{CO} & & \text{CO} \\
| & & | \\
\text{HCOH} \quad +\text{ATP} \longrightarrow & \text{HCOH} & +\text{ADP} \quad (11.3) \\
| & & | \\
\text{HCOH} & & \text{HCOH} \\
| & & | \\
\text{H}_2\text{CO}-\text{\textcircled{P}} & & \text{H}_2\text{CO}-\text{\textcircled{P}}
\end{array}$$

Ribulose-5-phosphate   Ribulose diphosphate

This completes the cycle; it assimilates one CO_2 in the form of glucose-6-phosphate at the expense of three ATP and two NADPH provided by photosynthetic activity. Because the only function of light is to provide the necessary ATP and NADPH, one could define photosynthesis as the set of light-driven reactions that generate these substances. The subsequent metabolism is common to photosynthetic and to some nonphotosynthetic organisms.

Our knowledge of the reductive pentose cycle did not spring fully developed like Athena from the head of Zeus; it represented a remarkable synthesis by Calvin and his collaborators of pieces of information developed earlier in many laboratories including Calvin's. The importance of this achievement is registered in Calvin's Nobel Prize in chemistry (1961) and in the fact that the reductive pentose cycle is usually called the Calvin cycle.

While Calvin's work was in progress, and before, several scientists including Fager, Gaffron, Gibbs, and Kandler had made experiments on $^{14}CO_2$ incorporation by illuminated algae. Also Horecker, Racker, and others were identifying some of the enzymes and partial reactions of the Calvin cycle in such "ordinary" bacteria as *Escherichia coli* (see a review by Vishniac et al., 1957, in the suggested reading). It is interesting to contrast the state of knowledge of $^{14}CO_2$ incorporation in 1954 with that in 1949 as expressed by A. A. Benson et al. and by E. W. Fager (suggested reading).

The demonstration of the Calvin cycle in isolated spinach chloroplasts had to await Arnon's success[2] in producing the conditions under which the chloroplasts would fix CO_2 at more than miniscule rates, about 1% of the rate in leaves. Rates approaching those in leaves were achieved with chloroplasts in about 1967.

The Calvin cycle is not merely responsible for carbohydrate synthesis; intermediates in the cycle are the starting points for the syntheses of fats and the amino acids that make up protein. The steady-state concentrations of intermediates are influenced by the activities of the 10 or more enzymes of the cycle, and the enzyme activities are in turn regulated by biochemical (feedback) control mechanisms. Thus the plant can divert more material toward carbohydrate, fat, or protein in response to the environment. By understanding these regulatory mechanisms[3] we can hope to improve agriculture, for example, by spraying crops with chemicals that lead to a greater proportion of protein. As an encouraging example, *Chlorella* cells respond to 10^{-3} M NH_4Cl by making far less sugar, less total carbohydrate, more fat, and far more amino acids.

The evolution of the Calvin cycle happened at a geological age when there was far more CO_2 in the atmosphere than there is today and far less oxygen. The crucial enzyme RuDP carboxylase is attuned to these former conditions; it has a rather low affinity for CO_2, and it catalyzes a seemingly aberrant reaction of O_2 with RuDP to form phosphoglycolic acid. Competing with the "normal" reaction RuDP + $CO_2 \longrightarrow$ 2PGA, the reaction with O_2 is

$$
\begin{array}{c}
H_2CO-\textcircled{P} \\
| \\
CO \\
| \\
HCOH \\
| \\
HCOH \\
| \\
H_2CO-\textcircled{P} \\
\text{RuDP}
\end{array}
+ O_2
\xrightarrow[\text{acting as oxygenase}]{\text{RuDP carboxylase}}
\begin{array}{c}
COOH \\
| \\
HCOH \\
| \\
H_2CO-\textcircled{P} \\
\text{PGA}
\end{array}
+
\begin{array}{c}
COOH \\
| \\
H_2CO-\textcircled{P} \\
\text{Phosphoglycolic acid}
\end{array}
\qquad (11.4)
$$

Some of the resulting glycolic acid is converted to serine and glyceric acid by way of glycine (the student can identify these molecules in any text on biochemistry), and CO_2 is released during this conversion. This process of oxygen consumption and CO_2 evolution is called photorespiration. It may help to replenish intermediates of the Calvin cycle, but at a price. Through photorespiration a plant can squander a large fraction of the CO_2 that it has assimilated by the "normal" operation of RuDP carboxylase. This waste is aggravated when the level of CO_2 is low and that of O_2 is high, and also at high temperature and light intensity. It is a very significant handicap for most contemporary plants when they encounter unusually hot sunny weather. The algae and plants brought this problem on themselves; through their profligate growth in earlier ages they put abundant O_2 into the atmosphere and removed most of the CO_2, storing it as fossil fuel and carbonaceous deposits. Now the atmospheric level of CO_2, 0.04%, is only a trace of what it once was.

The plants have responded in two ways that we shall examine in the next section. They have raised their content of RuDP carboxylase to an astonishing degree, and some species have developed an auxiliary metabolic cycle that concentrates CO_2 and delivers it to RuDP carboxylase.

11.2   The pyruvate carboxylation pathway

In 1965 it was discovered[4] that when $^{14}CO_2$ is given to sugar cane, the radioactivity appears in 4-carbon compounds such as malic acid before it appears in PGA. Further studies, especially by Hatch and Slack,[5] revealed the details of a CO_2-fixing mechanism common to tropical grasses (including sugar cane and corn) and found also in some unrelated plants. This "Hatch-Slack-Kortschak," or HSK, cycle starts with the conversion of pyruvic acid to a "high energy" phosphorylated enol form, phospho-enol pyruvic acid or PEP, and continues with the binding and later release of CO_2 in a cycle outlined in Fig. 11.2. At first sight this cycle appears to accomplish nothing, at the expense of the energy of two ATP (the $\Delta G_0'$ for conversion of pyruvic acid to PEP has the remarkably high value of nearly 15 kcal mole^{-1}). However, the high affinity of the CO_2-binding enzyme PEP carboxylase and the physical location of the cycle in the leaves of plants help to overcome the handicap incurred in using RuDP carboxylase in the present environment of low CO_2 and high O_2.

The HSK pathway is active in the mesophyll cells that are near the surfaces of a leaf, and the Calvin cycle activity is concentrated deeper in the leaf, in the vascular bundle sheath cells. Atmospheric CO_2 is fixed avidly by the PEP carboxylase in the mesophyll cells. It is transported in the form of malic acid (or a related amino acid in some species) to the bundle sheath cells by way of connecting filaments called plasmadesmata. There the CO_2 is released, becoming available to the Calvin cycle at a concentration more suited to the low CO_2 affinity of RuDP carboxylase. Furthermore any CO_2 released by photorespiration can be recaptured on its way out of the leaf and recycled through the HSK pathway. As a final, and perhaps more significant embellishment, there is evidence that the chloroplasts of the bundle sheath cells perform relatively more cyclic photophosphorylation and less noncyclic photometabolism, leaving most of the NADPH formation (and O_2 evolution) up to the mesophyll cells. The concentration of O_2 can thus be lower in the interior of the leaf, and less likely to drive the wasteful oxygenase activity of RuDP carboxylase. This picture is consistent with the fact that the longer wave light that drives cyclic phosphorylation by means of Photosystem 1 can penetrate more readily than shorter wave light to the interior of the leaf.

Plants that lack the HSK pathway expend a great amount of effort in overcoming their low affinity for CO_2, by the clumsy device of making extravagant amounts of RuDP carboxylase. This enzyme, comprising half the soluble protein of chloroplasts, is probably the most abundant enzyme on earth. Its concentration in the chloroplast is about 0.5 mM, comparable to the concentrations of its substrates. By keeping most of the RuDP in a bound form, the carboxylase may also prevent deleterious reactions of free RuDP. Being rather unstable, RuDP can react spontaneously to give compounds that inhibit the various metabolic systems.

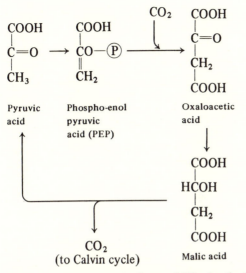

Fig. 11.2. Binding and release of CO_2 by the Hatch-Slack-Kortschak pathway. Details of phosphate and hydrogen transfer are omitted. This is apparently a mechanism for concentrating CO_2 and delivering it to the Calvin cycle.

The consequences of the HSK pathway can be summarized by saying that "C_4" plants, those that have this pathway, are well adapted to growth in bright sunlight and at high temperature (optimum about 35°C). The "C_3" plants that have only the Calvin cycle have lower CO_2 fixing efficiency under such conditions. They function optimally at about 20°C in light about one-tenth as intense as full sunlight. The difference can be illustrated dramatically by putting a corn plant and a bean plant under the same jar, exposed to air, strong light, and a relatively high temperature. CO_2 becomes transferred from the bean plant to the corn plant; the latter "eats" the former.

The Calvin cycle presents an interesting riddle in evolution. Its widespread occurrence suggests an extremely ancient origin, yet it remains the sole major mechanism for assimilating CO_2. Despite the difficulties imposed by a changing environment, no system has evolved to take its place; only auxiliary devices have evolved to help meet the new problems.

SUGGESTED READINGS

Bassham, J. A., and Calvin, M. (1957). *The Path of Carbon in Photosynthesis.* Prentice-Hall, Englewood Cliffs, New Jersey. 104 pp.

Bassham, J. A. (1965). Photosynthesis: The path of carbon. In *Plant Biochemistry*, J. Bonner and J. E. Varner, eds., pp. 875–902. Academic Press, New York.

Gibbs, M., and Kandler, O. (1957). Asymmetric distribution of C^{14} in sugars formed during photosynthesis. *Proc. Natl. Acad. Sci. U.S. 43*, 446–51.

Benson, A. A., Calvin, M., Haas, V. A., Aronoff, S., Hall, A. G., Bassham, J. A., and Weigl, J. W. (1949). C^{14} in photosynthesis. In *Photosynthesis in Plants*, J. Franck and W. E. Loomis, eds., pp. 381–401. Iowa State College Press, Ames, Iowa.

Fager, E. W. (1949). Investigation of the chemical properties of intermediates in photosynthesis. In *Photosynthesis in Plants*, J. Franck and W. E. Loomis, eds., pp. 423–36. Iowa State College Press, Ames, Iowa.

Vishniac, W., Horecker, B. L., and Ochoa, S. (1957). Enzymic aspects of photosynthesis. *Adv. Enzymol. 19*, 1–78.

Bassham, J. A., and Kirk, M. (1962). The effect of oxygen on the reduction of CO_2 to glycolic acid and other products during photosynthesis by *Chlorella*. *Biochem. Biophys. Res. Commun. 9*, 376–80.

Tolbert, N. E. (1974). Photorespiration. In *Algal Physiology*, W. D. P. Stewart, ed., pp. 474–504. Blackwells, Oxford.

Andrews, T. J., Lorimer, G. H., and Tolbert, N. E. (1973). Ribulose diphosphate oxygenase. I. Synthesis of phosphoglycolate by Fraction-1 protein of leaves. II. Further proof of reaction products and mechanism of action. *Biochemistry 12*, 11–18, 18–23.

Zelitch, I. (1971). *Photosynthesis, Photorespiration and Plant Productivity*. Academic Press, New York, 347 pp.

Zelitch, I. (1975). Pathways of carbon fixation in green plants. *Annu. Rev. Biochem. 44*, 123–45.

Fuller, R. C. (1978). Photosynthetic carbon metabolism in the green and purple bacteria. In *The Photosynthetic Bacteria*, R. K. Clayton and W. R. Sistrom, eds., Chapter 36, pp. 691–705. Plenum, New York.

Epilogue: Some directions of basic and applied research on photosynthesis

Twenty years ago one of the most enigmatic aspects of the photosynthetic mechanism appeared to be the primary conversion of quantum energy to chemical (redox) energy. We now have at least a sensible picture of this process, clothed in considerable detail in the case of the bacterial reaction centers. We have just begun to achieve the isolation of complexes approaching the status of reaction-center preparations from green plants, and the study of these materials, especially in their comparative aspects, will surely give us more insight into the primary physico-chemical mechanisms.

If we ask the simple question, "Why were chlorophylls chosen by primordial creatures as the photocatalysts of photosynthesis?" we can point to a combination of three virtues. First, the chlorophylls absorb light in the red and near infrared as well as in the blue, and with the help of accessory pigments they can receive energy from the entire spectrum of sunlight reaching the earth's surface. Second, chlorophylls do not squander the absorbed energy by extremely rapid radiationless de-excitation; this is evident from the high yield of Chl fluorescence in vitro. In contrast the iron-tetrapyrroles such as cytochromes show negligible fluorescence. The third virtue of chlorophylls for photosynthesis is their high versatility as electron donors and acceptors. Depending on their environment, they can operate around midpoint potentials ranging from $>+0.8$ V, for P/P^+ in Photosystem 2, to about -0.7 V for A_1/A_1^-, probably a specialized Chl a, in Photosystem 1.

Our conceptions of electron and H^+ transport driven by the reaction centers, connecting the primary redox energy to the useful products ATP and NADPH, are extensive but remain largely in the category of working hypotheses, subject to further change and elaboration. In this area the most effective tool has become the study of responses to brief flashes of light, and groups of flashes, at controlled redox potential and pH. It must be admitted that our opinions on this subject are based almost entirely on studies with subcellular preparations rather than with intact cells and living tissues.

The mechanisms by which H_2O is converted to O_2, and ADP to ATP, remain deep mysteries, but we may soon learn the chemistry of O_2 evolution. Our knowledge of molecular and supramolecular structure in photosynthetic membranes remains rudimentary. We are not yet able to see where in the thylakoid membrane (with all its bumpy contours) the Chl is actually located. We can only draw correlations between isolated Chl–protein complexes and particlelike features in the membrane. Through optical polarization methods we have gained information on the orientations of pigments, but we lack the translational coordinates that could be combined with the angular coordinates to give a three-dimensional structure. Detailed submolecular structures of the membrane proteins and protein complexes, including reaction centers, will come through X-ray diffraction analysis when someone has achieved the crystallization of these hydrophobic proteins.

In the practical realm we would like to improve agriculture and to use photosynthesis (natural, modified, or imitated) as a means of converting large amounts of solar energy into simple fuels and electricity.

An enormous saving in the energy cost of agriculture could be gained if plants could be taught to fix N_2. This might come about eventually through recombinant DNA techniques. We would hope that this would not be followed by an ecological catastrophe in which all the nitrogen in the atmosphere became converted to ammonia.

Organic matter can be fermented cheaply to give methane, and there are current explorations into the possibility of harvesting and cultivating some of the more abundant water-dwelling plants as sources of organic matter. Two such plants are the water hyacinth, in tropical waters, and kelp, in temperate seas. Coupled with this approach is a search for microbes (or their enzymes) that can digest cellulose rapidly. There are also some unusual plants that produce hydrocarbons; the economic aspects of their cultivation are being studied. Sugar from cane can be fermented to ethanol at a reasonable cost. All of these processes become more cost-competitive as the prices of conventional fuels rise.

Another possible approach is to modify photosynthesis so that the products are O_2 and H_2. This can be done with some algae by having an active hydrogenase that transfers electrons from the reducing side of Photosystem 1 to H^+; $2e^- + 2H^+ \xrightarrow{\text{Hase}} H_2$. There is a problem in that O_2 both inactivates the Hase and competes as an electron acceptor from Photosystem 1. This can be overcome by flushing the algal culture vigorously with N_2, so as to remove the O_2 as it is formed, but the H_2 must then be recovered from a gas that is mostly nitrogen. Another problem is that in most plants the electron transport chain, especially the plastoquinone pool, cannot keep pace with incoming quanta in full sunlight. The maximum rate of electron transport falls short by a factor of 10. There are some laboratory strains of *Chlorella*, however, that can keep up with full sunlight. And if the antenna can be made 10-fold smaller (by genetic manipulation), electron transport can keep pace with full sunlight in all cases. It is only necessary to compensate for the smaller antenna by using 10 times as much plant material, to assure adequate absorption of the solar flux.

A more exotic scheme is to make a photovoltaic cell, analogous to a silicon cell, from artificial chlorophyll-pheophytin complexes that behave like reaction centers. The most formidable problem in this approach is to achieve chemical stability in full sunlight over a period of years.

Some of these approaches to large-scale solar-energy conversion by modified or artificial photosynthesis are manifestly visionary, but perhaps no more so than the development of controlled nuclear fusion, and they are far less expensive to explore at the level of technical feasibility.

NOTES

Chapter 1

1. The notion that water is split in the photochemical act was retained for many years, apparently for no reason beyond tradition ("because it is there"). This has caused some semantic confusion because water splitting is also implicit in the peripheral reaction, $2H_2O \longrightarrow O_2$, when H_2O acts as the substrate H_2A. Contemporary mention of water splitting generally refers to this peripheral chemistry.

Chapter 2

1. It was for this that Einstein was awarded his sole Nobel Prize in 1921, three years after Planck's Nobel award.
2. A basic rule, the Franck-Condon principle, is that during the brief time of a transition, about 10^{-15} sec for transitions involving visible light, the nuclei do not have time to change their positions.
3. Certain bacteria, exemplified by *Halobacterium halobium*, contain no chlorophyllous pigments, but their cytoplasmic membranes develop purple patches when the bacteria are grown in strong light. The purple pigment is bacteriorhodopsin, a close analog of the visual pigment in the eyes of animals. In the purple patches this pigment mediates a light-dependent formation of ATP. The mechanism is different from that of photosynthesis based on Chl, and photometabolism does not usually play a major part in the life of these bacteria. Their properties have been described by W. Stoeckenius; see the suggested reading for this chapter. We shall return to this singular case of photosynthesis without chlorophyll in Section 6.4.
4. The reaction center is not strictly an irreversible trap; a quantum of excitation in the reaction center has a finite probability of returning to the antenna instead of being used for photochemistry. Evidence from some photosynthetic bacteria suggests a rate constant less than 10^{10} sec^{-1} for transfer out of the reaction center, and a rate constant greater than 10^{11} sec^{-1} for photochemical utilization. These reaction centers are thus "better than 90%" irreversible. It remains possible that in other organisms the rate constant for transfer out of the reaction center is much greater, comparable to that for photochemistry. Then a quantum may encounter several reaction centers before being "trapped" in one by photochemical utilization. This would not affect the concept of "transfer time," the average lifetime

of an excitation quantum before it is trapped. The reaction centers would simply be acting like antenna pigments part of the time.

5. In some photosynthetic bacteria there are two antenna components, each containing Bchl *a* in a particular environment. The Bchl *a* has its long wave absorption maximum at 850 nm in one component and at 875 nm in the other. These antenna systems are coupled to reaction centers in which the long wave absorption maximum is at 870 nm.

6. The ability of photochemically inactive reaction centers to dispose of excitation energy may be a useful safety valve against the accumulation of potentially harmful amounts of energy in the system.

7. This was shown by L. N. M. Duysens in a doctoral thesis that stands as a landmark in the literature of photosynthesis. Studying energy transfer in photosynthetic bacteria, Duysens measured the relative effectiveness of light absorbed by carotenoids and by Bchl, in promoting photosynthesis and in eliciting fluorescence of the antenna Bchl. The relative efficiencies were the same for both phenomena in a variety of examples. It could be concluded that energy absorbed by the carotenoids is first transferred to the Bchl, with limited efficiency. Excitation energy in the Bchl, whether absorbed directly or received from the carotenoids, is used mainly for photosynthesis, but a small fraction is given up as fluorescence.

8. The terms "blue-green" and "red" in describing these algae arose because phycocyanin, a blue phycobilin, is predominant in the former, and the red phycoerythrin is the major phycobilin in the latter. In some algae the distinction is blurred because they synthesize relatively more phycoerythrin when grown in green ($\sim$570 nm) light and more phycocyanin when grown in red-orange ($\sim$630 nm) light. This physiological response is known as chromatic adaptation.

Chapter 3

1. A method developed by P. Joliot uses modulated light and a rapidly responding oxygen electrode. Photosynthetic O_2 evolution is modulated in synchrony with the light, with a phase shift of 1 msec reflecting the delay between the absorption of light by the algae and the appearance of O_2 at the electrode. Respiration appears as an unmodulated uptake of O_2.

2. In modern terms these reserves are quinones that can accept electrons from Photosystem 2, allowing O_2 evolution to be driven for a short time by Photosystem 2 alone, without the cooperation of Photosystem 1.

3. The disproportionate effectiveness of phycobilins for photosynthesis could already have been discerned in the experiment made by Engelmann (Section 2.6) 70 years earlier.

Chapter 4

1. Rotational motion can be assigned a direction that corresponds to the advance of a right-handed screw.

2. Individual spins must actually be added as vectors of magnitude $[s(s + 1)]^{1/2}$, where s is the spin quantum number, $\frac{1}{2}$ for a single electron. The quantity $(h/2\pi)[\frac{1}{2}(\frac{1}{2} + 1)]^{1/2}$ is the angular momentum of an electron due to its spin. For antiparallel electron spins the vectors are equal and opposite, and their sum is zero ($S = 0$). For two "parallel" spins, where the sum is $S = \frac{1}{2} + \frac{1}{2} = 1$, the resultant of the vector addition must have the magnitude $[1(1 + 1)]^{1/2}$:

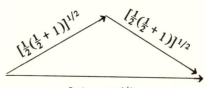

$$[1(1 + 1)]^{1/2}$$

The two spins are not actually parallel; each one makes an angle of $35°$ with the resultant, as one can compute from the above diagram. The expression "parallel" is used merely for convenience.
3. The spectroscopists of the nineteenth century classified sets of lines in the hydrogen spectrum as s, p, and d for "sharp," "principal," and "diffuse."
4. More correctly, "concentration" should be replaced by "activity," an analogous parameter in which departures from the behavior of ideal solutions have been taken into account. We shall use the less precise but more familiar term.
5. It is beyond the scope of this book to develop the distinction between total energy and free energy (available for performing work) in terms of the Second Law of Thermodynamics. The energies described by redox potentials are free energies.
6. Absorbance (a) is beginning to replace the synonymous optical density (OD) in the photobiological literature. It is often symbolized A, but we shall reserve A to denote the fraction of incident light absorbed by a sample.
7. For Chl a, ϵ is about 100 mM^{-1} cm^{-1} at the maximum of absorption. Inspection of the absorption spectrum suggests that an average value of ϵ, taken over the spectrum of sunlight reaching the earth's surface, is about one-twentieth the peak value or about 5 mM^{-1} cm^{-1}. The average absorption cross section α is about 0.2×10^{-16} cm^2 per molecule. The intensity of full sunlight at the zenith is about 0.1 W cm^{-2}, and with a median wavelength of 600 nm this corresponds to 30×10^{16} quanta cm^{-2} sec^{-1}. The product αI is then $(0.2 \times 10^{-16})(30 \times 10^{16})$ or 6 quanta sec^{-1}; this is roughly the rate at which a single molecule of Chl a absorbs quanta in direct sunlight with the sun at its zenith.
8. This procedure can sometimes be facilitated by interpolation between measured responses that are slightly larger and smaller than the standard one.

Chapter 5

1. A simple way to isolate reaction centers from carotenoidless mutant *Rp. sphaeroides* is to break the cells, isolate the chromatophores by centrifugation, expose the resuspended chromatophores to the detergent lauryl dimethyl amine oxide (LDAO):

$$C_{12}H_{25} - \overset{\overset{\displaystyle CH_3}{|}}{\underset{\underset{\displaystyle CH_3}{|}}{N^+}} - O^-$$

then remove contaminating material by centrifugation, and purify the reaction centers by fractional precipitation with ammonium sulfate. Other useful techniques for purifying reaction centers are chromatography on diethylaminoethyl cellulose or on hydroxylapatite; the latter method often succeeds where other methods fail.

2. Bpheo is Bchl in which the central magnesium atom has been replaced by two H atoms.

3. Molecular weights can be estimated by electrophoresis through polyacrylamide, after dissociating the protein into its component polypeptides by heating with the detergent sodium dodecyl sulfate (SDS) in a reducing environment. This technique of SDS-PAGE (polyacrylamide gel electrophoresis using SDS) is of great value in biochemistry because the electrophoretic mobilities of polypeptides are determined, to a first approximation, not by their intrinsic charge but by their coating of negatively charged SDS. As a result their electrophoretic mobilities are related simply to their molecular weights; these can be found by comparison with the mobilities of polypeptides of known molecular weight. The "marker proteins" of known molecular weight are all hydrophilic, and the molecular weights inferred for hydrophobic polypeptides should be taken with caution lest their mobilities differ anomalously from those of the markers. There is growing evidence that hydrophobic proteins have greater affinity than hydrophilic ones for SDS, causing the molecular weights of the former to be underestimated by about 30%. Even so, SDS-PAGE has been a major tool for characterizing the hydrophobic proteins associated with membranes in biological structures.

4. 980 nm in cells and chromatophores; 960 nm in isolated reaction centers.

5. Radiationless de-excitation can proceed by more than one mechanism. In Chls *a* and *b* the main pathway is conversion from the excited singlet state to a triplet state (*intersystem crossing*), followed by decay of the triplet state to the ground (singlet) state. In Bchl *a* another mechanism (*internal conversion*) may be predominant: The energy of the excited singlet state is dissipated as heat by a sequence of small transitions through vibrational and rotational sublevels of the excited and ground singlet electronic states.

6. These are single mutations, as judged by reversion frequencies of the order of 10^{-7}.

Chapter 6

1. The use of spinach as a source of chloroplasts has been so predominant, almost universal, that the reader can assume that all remarks about chloroplasts in this book pertain to spinach unless otherwise stated.
2. Such fractions cannot evolve O_2, but the natural flow of electrons from water can be replaced by an artificial electron donor such as hydroquinone. The resulting Photosystem 2 activity can be distinguished from Photosystem 1 activity by details of the light-induced changes in the reaction centers and by the distinctive effects of certain inhibitors.
3. These apparent weights, based on electrophoresis in polyacrylamide, vary considerably in the reports of different investigators. They are based on electrophoretic mobility, which varies significantly with changes of methodological details.
4. K. Satoh (1979), *Biochim. Biophys. Acta 546*, 84–92.
5. These sizes are not well defined. In a histogram showing the relative number of bumps of different sizes there are two bands with maxima at 8 and 11 nm, but the width of each band at half-maximum is about 5 nm.
6. The fluorescence bands at 680, 685, and 735 nm were seen in fractions isolated by the use of digitonin. Stronger detergents lead to fluorescence at shorter wavelengths, probably by disengaging some of the Chl from its native environment. This could account for the discrepancy between 695 nm fluorescence in the leaf and 685 nm fluorescence in Fraction 2.
7. The German word *purpur* means either purple or dark red.
8. A testimony to the decline of the arts of sterile microbial methodology. Many who work with photosynthetic bacteria do not even own a microscope.

Chapter 7

1. The principal bands are modulated, giving secondary bands, by nuclear vibrations.
2. The energy flux is given by the Poynting vector $\overline{E} \times \overline{B}$, and the magnitude of the magnetic vector B is proportional to E.
3. For example, a Chl dimer has maximum overall dimensions about 5 nm.
4. For n interacting dipoles there are n independent modes with resultant transition moments given by appropriate linear combinations of $\overline{M}_1, \overline{M}_2, \ldots, \overline{M}_n$, each multiplied by a suitable positive or negative constant.
5. Even a partial orientation can be useful. In a crystal the unit cell may contain several molecules, each with different orientations, but this set of orientations is repeated in every unit cell. A flat piece of photosynthetic membrane, dried onto a glass plate, can have molecules that are oriented randomly with respect to rotation about an axis perpendicular to the plane of the plate, but uniformly with respect to the inclinations of molecular axes from the plane.
6. The extent of rotational depolarization can be represented in an

artificial but mathematically useful way, by imagining that at any instant (at a time t after the photoselective excitation) a fraction f of the photoselected population of molecules has become entirely randomized, and the remainder $1 - f$ has retained its initial orientation. The actual population, of course, does not behave in this way, but the half-time can be taken as the time at which the randomization is equivalent to $f = 50\%$.

7. The angle of refraction can be computed from the angle of incidence if the index of refraction is known. The latter can be measured by focusing a microscope (through a known distance) onto the upper and lower surfaces of the sample, or by measuring absorbance with the electric vector aligned with the XY plane and the axis of propagation first perpendicular to the plane and then making a known angle with it.

8. C. N. Rafferty, J. Bolt, K. Sauer, and R. K. Clayton (1979), *Proc. Natl. Acad. Sci. U.S. 76*, 4429-32.

9. M. S. Davis, A. Forman, and J. Fajer (1979), *Proc. Natl. Acad. Sci. U.S. 76*, 4170-4.

10. The origin of this and other band shifts in reaction centers is obscure. For the most part the absorption bands of Bpheo are shifted to greater wavelengths when the quinone is reduced, but at low temperature the shift of the 547 nm band is reversed to a blue-shift. Various shifts of the bands of Bchl also attend the photochemistry. These shifts are regarded vaguely as "electrochromic," analogous to the atomic Stark effect in which energy levels are perturbed by electric fields. In reaction centers the local electric fields are presumed to interact with permanent and inducible dipoles in ground and excited states of the pigments. The various patterns of these shifts, in response to transient charged entities in the reaction center, may eventually help to elucidate structural relationships. They will be considered again in Section 8.1 and in Part IV.

11. K. D. Philipson, V. L. Sato, and K. Sauer (1972), *Biochemistry 11*, 4591-5.

12. These transitions are analogous to the 1250 nm transition of the oxidized special pair of Bchl shown in Fig. 5.11b. They are expected from theory to be parallel to the long wave Q_y transition of the corresponding special pair in its neutral state (or coplanar in the case of circular degeneracy).

13. J. Biggins and J. Svejkovsky (1978), *FEBS Lett. 89*, 201-4.

14. Crude diffraction patterns, conveying little information, have been obtained from the partial order of reaction centers in phospholipid multilayers, and more information may be forthcoming from electron microscopy of such preparations. Electron micrographs of repetitive structures can be subjected to the technique of optical diffraction, followed by processing of the diffraction pattern and reconstruction of an image in which the repeating features have been enhanced. This method has been applied recently to the surfaces of photosynthetic membranes, revealing crystal-like order, but not yet to ordered arrays of reaction centers.

15. These films, which resemble Kodak-Wratten optical filters, retain their excellent optical clarity at temperatures down to 4 K. They are ideal for studies at low temperature.

16. This is the only context in which spherical trigonometry has proved useful in the study of photosynthesis.
17. R. K. Clayton, C. N. Rafferty, and A. Vermeglio (1979), *Biochim. Biophys. Acta 545*, 58–68; C. N. Rafferty and R. K. Clayton (1979), *Biochim. Biophys. Acta 546*, 189–206.
18. A Vermeglio, J. Breton, G. Paillotin, and R. Cogdell (1978), *Biochim. Biophys. Acta 501*, 514–30.
19. M. J. Pellin, K. J. Kaufmann, and M. R. Wasielewski (1979), *Nature (London) 278*, 54–5.

Chapter 8

1. For a modern quantum-mechanical treatment of vibronically assisted electron tunneling, see J. Jortner in the suggested reading for this chapter.

2. Ortho-phenanthroline, or 1,10-phenanthroline, is an iron-binding (chelating) agent that appears to block the flow of electrons from Q_a to secondary quinones in reaction centers and membranes of photosynthetic bacteria.

3. In photosynthetic membranes, with a quantum of excitation energy able to visit more than one reaction center ("lake" model; see Section 2.2), the photochemical efficiency is not a linear function of the fraction of active reaction centers. Nevertheless, the efficiency of photochemical electron transfer remains proportional to $f_m - f$, barring other complications.

4. Mössbauer spectroscopy measures the gamma rays emitted and absorbed during transitions between energy states of radioactive nuclei such as ^{57}Fe. These energy levels are so sharply defined that resonant emission and absorption between nuclei is spoiled by the Doppler effect of relative movements of the nuclei as slow as 1 cm sec^{-1}. This permits one to analyze the recoil of a nucleus engendered by a transition, and hence to learn something about the surroundings (electrons and other nuclei) of the recoiling nucleus.

5. The iron-binding reagent o-phenanthroline inhibits electron transfer from Q_a^- to Q_b, but the mechanism of this inhibition remains puzzling. The absorption of X-rays by iron, using the technique of EXAFS (extended X-ray absorption fine structure), is a sensitive probe of the electronic environment of the iron. Application of this technique indicates that o-phenanthroline is not bound to the iron in reaction centers.

6. If cells are treated to remove the cell wall, the resulting spheroplasts have outer surfaces that correspond to the inner surfaces of chromatophore membranes.

7. Artifacts could include energy transfer among excited singlet states and multi-photon interactions. The early absorbance transients vary conspicuously, depending on the wavelength and intensity of excitation.

8. The theory of this "chemically induced dynamic electron polarization" (CIDEP) is beyond the scope of this book. For a qualitative description see J. J. Katz et al. (1979) in the suggested reading for this chapter.

9. For simplicity, and because the participation of Pheo a as an electron carrier between P and Q_a has not been proved, PIQ_a is written PQ_a here and in the ensuing formulas.

Chapter 9

1. F. L. Allen and J. Franck (1955), Photosynthetic evolution of oxygen by flashes of light, *Arch. Biochem. 58*, 124–43. In the earlier experiments of Emerson and Arnold (suggested reading, Chapter 2) the summed yield of many flashes was measured.
2. Originally Joliot measured the initial rate of O_2 evolution in weak continuous light after n conditioning flashes and found this rate to be maximal for $n = 2, 6, 10, \ldots$ Later, in collaboration with B. Kok, he measured the O_2 elicited by each flash. The yield of flash $n + 1$ was equivalent to the rate after n conditioning flashes. Experiments with spinach chloroplasts gave essentially the same results as those with *Chlorella* cells.
3. T. Yamashita and W. L. Butler (1968), *Plant Physiology 43*, 1978–86.
4. The possibility of electron donation to P^+ has provided a valuable assay for Photosystem 2 reaction centers in chloroplasts and subchloroplast fractions that have lost their ability to evolve O_2. Effective donors include hydroquinone (HO—Ph—OH), p-phenylenediamine (H_2N—Ph—NH_2), and 1,5-diphenylcarbohydrazide (Ph—$(NH)_2CO(NH)_2$—Ph).
5. M. Spector and G. D. Winget (1980), *Proc. Natl. Acad. Sci. U.S. 77*, 957–9.
6. DCPIP was first used in photosynthesis research by C. S. French and A. S. Holt at the University of Minnesota in 1941.
7. R. P. Levine (1969), *Annu. Rev. Plant Physiol. 20*, 523–40.
8. Levine did not actually observe the oxidation and reduction of PCy directly. The flow of electrons through PCy to P700 was inferred from observations of $NADP^+$ reduction driven by Photosystem 1, with reduced DCPIP serving as an external electron donor. This donor cannot pass electrons to P700 unless PCy is present as a mediator. Electron flow from DCPIP to $NADP^+$ was seen in mutants lacking Cyt f, but not in those lacking PCy. Intactness of the pathway from Photosystem 2 to Photosystem 1 (through components including Cyt f and PCy) was tested by measuring the reduction of $NADP^+$ with Photosystem 2 (drawing upon H_2O) as the source of electrons.
9. R. Malkin and P. J. Aparicio (1975), *Biochem. Biophys. Res. Commun. 63*, 1157–60. Also see J. Whitmarsh and W. A. Cramer (1979), *Proc. Natl. Acad. Sci. U.S. 76*, 4417–20.
10. D. I. Arnon and R. K. Chain (1975), *Proc. Natl. Acad. Sci. U.S. 72*, 4961–5.
11. Cytochromes of both the b and c types contain heme, a tetrapyrrole with Fe at the center, associated with protein. They differ in the way that the heme is bound to the protein, covalently in Cyt c and noncovalently in Cyt b. The tetrapyrrole framework in Cyt b has

two peripheral vinyl ($-CH=CH_2$) groups; in Cyt c these groups are reduced and linked covalently to cysteine residues in the protein:

$$Cys-SH + -CH=CH_2 \longrightarrow \quad \begin{array}{c} \diagdown \\ Cys-S \end{array} CH-CH_3$$

As a rule the c-type cytochromes have E_m values more positive than the b types, and absorption maxima of the α band at shorter wavelengths, near 550 nm for c and near 560 nm for b. There are many exceptions to this rule.

12. The high energy state is identified (see Chapter 10) with the electrochemical gradient that results from the transport of H^+ across the membrane:

$$\left. \begin{array}{l} \text{Electron transport,} \\ H^+ \text{ translocation} \end{array} \right\} \longrightarrow \ominus \longrightarrow \text{ATP formation}$$

13. A Vermeglio (1977), *Biochim. Biophys. Acta 459*, 516–24; C. A. Wraight, *Biochim. Biophys. Acta 459*, 525–31.

Chapter 10

1. This difference between the behaviors of chloroplast thylakoids and bacterial chromatophores may have as much to do with the relative sizes of the vesicles, and the resulting differences in surface/volume ratios, as with the relative permeabilities of the membranes; see the next section.

2. Another method, extremely sensitive and convenient, is to mix the sample with an extract of firefly tails and measure the equivalent emission of light. The luciferin–luciferase bioluminescent system of the firefly requires ATP, and the amount of light emitted is proportional to the amount of ATP available. In using this method one must take care that neither the sample nor the firefly extract contains an appreciable activity of adenylate kinase, which catalyzes the reaction $2ADP \longleftrightarrow ATP + AMP$.

3. A. T. Jagendorf and E. Uribe (1966), *Proc. Natl. Acad. Sci. U.S. 55*, 170–7.

4. P. Gräber, E. S. Schlodder, and H. T. Witt (1977), *Biochim. Biophys. Acta 461*, 426–40.

5. This buffering greatly increases the storage capacity of the "proton battery."

6. B. Chance and L. Smith (1955), *Nature (London) 175*, 803–6.

7. J. B. Jackson and A. R. Crofts (1969), *FEBS Letters 4*, 185–9.

8. B. L. Strehler and W. Arnold (1951), *J. Gen. Physiol. 34*, 809–20.

9. A. R. Crofts, C. A. Wraight, and D. E. Fleischman (1971), *FEBS Letters 15*, 89–100.

10. Actually a quantity somewhat less than $\Delta\psi$ should be used in this formula, because the electron does not span the entire thickness of the membrane in moving from P to Q_a. If the distance from P to Q_a projected on the perpendicular to the membrane is d, and the membrane potential spans a thickness D, the quantity $\Delta\psi$ should be

multiplied by d/D. In principle this provides a way to measure d/D; this has not yet been tried.

11. C. A. Wraight and A. R. Crofts (1971), *Eur. J. Biochem. 19*, 386-98.
12. W. Junge, B. Rumberg, and H. Schröder (1970), *Eur. J. Biochem. 14*, 575-81; H. Schröder, H. Muhler, and B. Rumberg (1972), *Proc. 2d International Congress on Photosynthesis*, G. Forti, M. Avron, and A. Melandri, eds. (Dr. W. Junk N. V. Publishers, The Hague), pp. 919-30.
13. A. R. Portis, Jr. and R. E. McCarty (1974), *J. Biol. Chem. 249*, 6250-4.
14. J. B. Jackson, S. Saphon, and H. T. Witt (1975), *Biochim. Biophys. Acta 408*, 83-92; K. M. Petty and J. B. Jackson, *FEBS Letters 97*, 367-72.
15. Recall that activity is the equivalent of concentration with departures from the behavior of ideal solutions taken into account. In most biochemical applications the activity is nearly equal to the concentration.
16. I. J. Ryrie and A. T. Jagendorf (1972), *J. Biol. Chem. 247*, 4453-9.
17. Actually three ATP are needed to fix one CO_2; the deficit may be supplied by Photosystem 1 through cyclic phosphorylation. The reader can derive an amended budget on the basis that one quantum can provide the missing $\frac{1}{3}$ ATP.

Chapter 11

1. J. A. Bassham, A. A. Benson, L. D. Kay, A. Z. Harris, A. T. Wilson, and M. Calvin (1954), *J. Am. Chem. Soc. 76*, 1760-70.
2. D. I. Arnon (1955), *Science 122*, 9-16.
3. J. A. Bassham (1971), *Proc. Natl. Acad. Sci. U.S. 68*, 2877-82; also J. A. Bassham (1971), *Science 172*, 526-34.
4. H. P. Kortschak, C. E. Hartt, and G. O. Burr (1965), *Plant Physiol. 40*, 209-13.
5. M. D. Hatch and C. R. Slack (1966), *Biochem. J. 101*, 103-11; also (1970), *Annu. Rev. Plant Physiol. 21*, 141-62.

INDEX

absorption coefficient, 72
absorption spectrum, 23-6
Acker, S., 186
action spectrum, 77-8
adenosine diphosphate (ADP), 231
    structure, 15
adenosine triphosphate (ATP), 9, 12, 231
    electron-proton stoichiometry in forma-
        tion of, 248-50
    firefly assay, 246, 274
    free energy of hydrolysis, 249-50
    structure, 15
allophycocyanin, 126
allophycocyanin B, 126
δ-aminolevulinic acid, 32, 108
    structure, 32
Anderson, J. M., 111
antenna (light harvesting), 27-8
    of green plants, 112-14, 118, 122-6
    physiological regulation, 106-8
    puddle and lake models, 30-1, 104
    tripartite organization in plants, 113,
        115, 123
    varieties in bacteria, 105-6, 108-9
Aparicio, P. J., 209
Armond, P., 81
Arnold, W., 28-30, 246
Arnon, D. I., 210, 258
Arntzen, C. J., 80, 81, 117
ATPase, see coupling factor
Austin, L. A., 151, 155

B850, 108-9, 152
B860, 108-9
B875, 108-9
bacteriochlorophyll a
    absorption in vivo vs. in vitro, 149-52
    absorption maxima, 38
    absorption spectra, 147
    biosynthesis, 32
    structure, 14

bacteriochlorophyll b
    absorption maxima, 38
    biosynthesis, 32
bacteriochlorophyll c, 127
    absorption maxima, 38
bacteriopheophytin
    absorption band shifts, 170
    in reaction centers, 154
bacteriorhodopsin, 129
    and proton transfer, 129-30
Barber, J., 122
Bassham, J. A., 257
Beer's Law, 72
Bendall, F., 56
Bengis, C., 113-14, 186
Benson, A. A., 258
Blinks, L. R., 53, 54
Boardman, N. K., 111
Bohr, N., 21-2
Boltzmann constant, 42
Boltzmann factor, 42, 247
Bonaventura, C., 120
bundle sheath cell, 260
Butler, W. L., 123-5

C550, 187, 189
Calvin, M., 256-8
Calvin cycle, 256-9
carbonylcyanide m-chlorophenylhydra-
    zone (CCCP), 244
Cario, G., 49
β-carotene, structure, 35
carotenoids
    absorption band shifts, 241-6
    absorption spectra, 39
    as antenna pigments, 36
    protective action, 34-6
Chain, R. K., 210
chemical coupling hypothesis, 232-4
chemiosmotic hypothesis, 234-5, 238
chirality, 135